Embrace Hardship To Forge Your Warrior Spirit

Ricardo Perez

As a young boy, I had to lecture my father about the dangers of shooting at people after he fired upon a vehicle that had just dropped me off at home. Memories have a funny way of losing details with time, but I consulted with family, friends, and those I served with to ensure accuracy in what I lay out. The names of the people in this story have been changed, but the craziness that ensues is the complete truth as I know it. It's an ugly truth, and it's time we face it down for what it is so we can begin fixing the wrongs and start the healing process. It's been far too long...

Born To Fail
Embrace Hardship To Forge Your Warrior Spirit
Editorial Work By: Anna Ciummo
ISBN 979-8-9851053-0-8 Hardcover
979-8-9851053-1-5 Paperback
979-8-9851053-2-2 Audiobook
979-8-9851053-3-9 Ebook

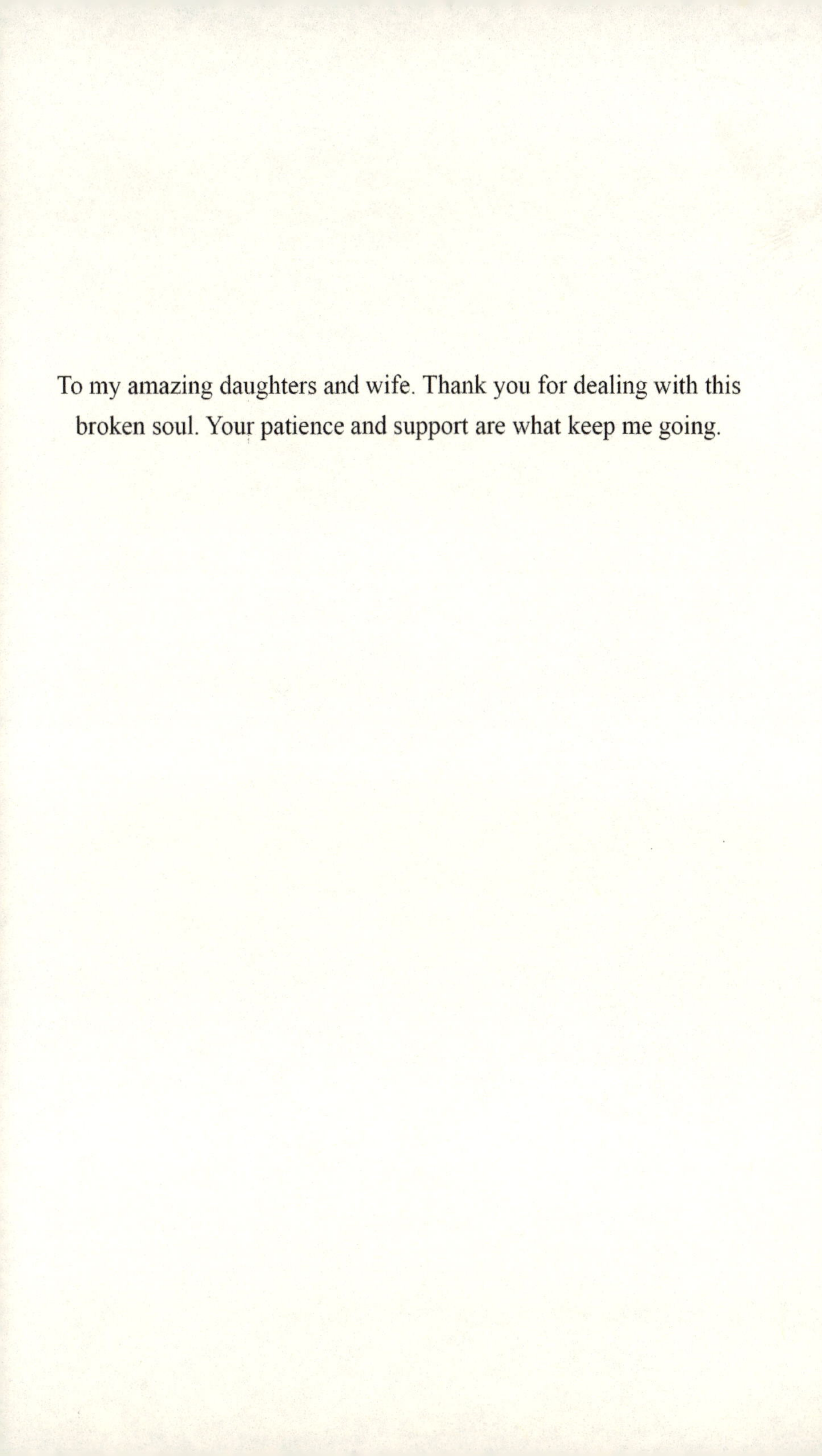

To my amazing daughters and wife. Thank you for dealing with this broken soul. Your patience and support are what keep me going.

Contents

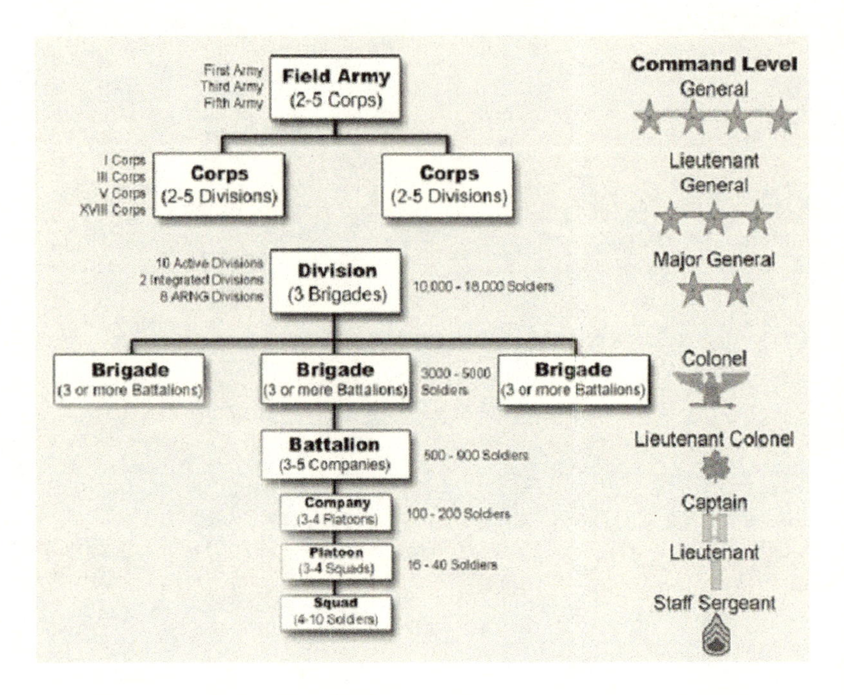
First Army
Third Army
Fifth Army
Field Army
(2-5 Corps)
I Corps
III Corps
V Corps
XVIII Corps
Corps
(2-5 Divisions)
Corps
(2-5 Divisions)
10 Active Divisions
2 Integrated Divisions
8 ARNG Divisions
Division
(3 Brigades)
10,000 - 18,000 Soldiers
Brigade
(3 or more Battalions)
Brigade
(3 or more Battalions)
3000 - 5000
Soldiers
Brigade
(3 or more Battalions)
Battalion
(3-5 Companies)
500 - 900 Soldiers
Company
(3-4 Platoons)
100 - 200 Soldiers
Platoon
(3-4 Squads)
16 - 40 Soldiers
Squad
(4-10 Soldiers)
Command Level
General
Lieutenant
General
Major General
Colonel
Lieutenant Colonel
Captain
Lieutenant
Staff Sergeant

Introduction

I started writing this in 2009 while deployed to the Arghandab River valley of Southern Afghanistan, and I served in the 5th Brigade, 2nd Infantry Division (or 5/2). We deployed to this region as part of President Obama's troop surge for the country and were the first American presence in the area for at least three years. A Canadian unit consisting of a few hundred men was previously tasked to keep an estimated 15,000 Taliban fighters from capturing Kandahar, Afghanistan's 2nd largest city. The Canadian force lost 125 soldiers during that time while killing thousands of Taliban fighters. That deployment was especially tough for the men in the 1st Battalion, 17th Infantry Regiment, one of five battalions made up 5/2.

I served in the Battalions Alpha Company (A.co) from its inception. I was one of the first soldiers in 5/2 when the Army put me there in 2006, and it was that wild ride we went on that bound us together. We were less than 800 men in charge of the entire Arghandab area, and as a result, twenty-three men lost their lives, making it one of the highest death tolls of any U.S. Army Battalion deployed to Afghanistan. I was in a lot of pain, physically and mentally, and couldn't believe the things I was experiencing daily, so I began to document everything I could. Since then, I've come a long way in personal and professional development, but I feel like the world is falling apart with everyone losing their minds, and telling me I'm the crazy one. People have been given new platforms that create feedback loops, reinforcing their mental states, and software engineers have developed sophisticated algorithms that can give us more of what IT thinks we want.

That means if you're regularly depressed and choose to listen or watch depressing shit, you're going to be recommended or fed more depressing shit by the platforms. I think that is why the rates of depression, mental illness, and suicide are at an all-time high, with the rest who barely keep it together only able to do so with the help of pharmaceuticals. It's easier now than ever to distract ourselves, and there are endless reasons not to sit down and wrestle with our thoughts or be self-critical. There have always been small things that distract us, but the world has changed so drastically in such a short amount of time that most people don't truly realize the implications. People used to regularly reflect on themselves and their worldviews and disagree about stuff without their logic breaking down. Now they just put blinders on, get on social media, and spread their or others' "expertise" across the digital landscape.

The amount of misinformation people is exposed to truly frightens me. We live in a unique time where these platforms have given malicious entities a direct line to feed us propaganda.

I don't mean that term in a political sense; there is a wide spectrum of topics that have been used to perpetuate mistrust throughout our country. Growing up, I wasn't able to distract myself much; my life was one of darkness and terror. The parental support I got came by default; they did just enough to keep themselves out of jail. There were no annual check-ups, concern for academics, or any guidance other than "Join the military." To make things worse, I tried to talk as little as possible.

"Shy" or "Shy Boy" were the nicknames my best friends in high school gave me to give you an idea of what level I took that to. I raised myself from an early age, and the voyage we're about to go on will show you what I think is the most effective way to change your reality.

It's hard, and will be the hardest thing you'll ever do if you decide to embark on the journey, but once you reach your first milestone, the path to where you want to go becomes much clearer. Life dealt me a bad hand that forced me to baste in my fluids, so I'm familiar with failure and wanting to quit. I went from a broken home to a broken military unit, and the gauntlet that's been my path has been fraught with hardship and turmoil. I let self-doubt stop me when writing this book, and even though I did, I knew I had to keep myself engaged with something, or I would end up depressed and spiteful.

You have to recognize your triggers. I drugged myself from a place of darkness, and I was going to do everything in my power to keep myself from falling back in. I was trapped in my mind my whole life, but the military showed me that I'm the 0.1%, and with that, the mental cage I had been in broke down. I finally became aware of what I, and most of us, are capable of. I stopped making excuses and focused strictly on doing whatever I had to do to take on the challenges presented to me. It's the secret to bringing your dreams closer to reality, and I've been using it as a method to achieve things I would've never thought possible. Stop buying bullshit and use that money to invest in yourself because those material things are what's making you soft.

One of my favorite quotes is "If you think education is expensive, try ignorance." Eppie Lederer wrote this in her column "Ask Ann

Landers" in response to a question she received from parents wanting tax deductions on paying tuition to educate their four children. Get uncomfortable. Complacency is the shackle that holds us down, and it's easier than ever to sit on our thumbs.

Failure and self-reflection are the most powerful weapons in our self-development arsenal, and you've been shooting blanks for years. Fear of failure is the most crippling thing we do to ourselves, and what you probably don't realize is that the greatest minds fail every day. They go to their labs to try an experiment, and it usually doesn't give them their desired results. But they analyze, reflect, and gain more knowledge about the problem they're trying to solve and themselves with each iteration. Nothing worth doing is easy or comes fast, and if it does, chances are you'll be ill-prepared to handle everything.

Through my trials and tribulations, I've become a rockstar without fortune, fame, or talent, so when my day comes, I'll be ready to take on any new challenges. I'll give it my best, probably fuck something up, but will always, as they say, "fail forward." Once you've set your mind free by finding your true self, you'll be able to make your visions a reality, and I hope my journey can help illuminate your path to self-discovery.

CHAPTER ONE

BACK FOR ANOTHER TRY

It was a dry and frigid winter—the kind where air consumed all heat from the small gas-heated house. I was the second born in the family and made my journey from the womb into an environment unprepared for a kid.

I learned early on that the pregnancy was an accident. And that was how I learned to think of myself since the beginning: an accident. The default parental support I got didn't help my situation either. As a kid my parents never took me to the doctor, had concerns for my academics, or gave me any true life advice other than things like "just join the military." My limited life made me feel small, which is how I learned to talk as little as possible and hide myself away, never even considering that I could be better. "Shy or Shy Boy" was the most common nickname I was given by my friends in high school.

I don't remember a lot of my young life; happy memories are few and far between, and most of the things I do remember are unfortunately bleak. Even from a young age, I was already trying to find ways to distract myself from my own small reality.

The best distractions I had as a kid were guns, video games, and fights with my family and the kids at school. Some of these moments of distraction I even look back on with fondness. In fact, one of the best memories from my youth was my time wandering around my family's ranch with my One-Pump BB gun. In grade school, my dad would often take me hunting for rabbits or shooting on the side of the ranch we didn't live on. Though that was the extent of our father-son bonding time, they're still great moments that I cherish. I just wish there could've been thousands of other moments like that.

Because Dad wasn't always around, I spent many of my young years unsupervised, walking around my ranch with my BB gun. I would often get my mind stuck on hunting a particular animal.

My dad's brother Ruperto had a hunting lease, and the only deer I ever killed happened when I was about eight years old. I remember it being early in the morning, the sun just peeking through, but still somehow blinding as we sat in our deer blind. About 150 yards down the dirt road, a doe and her fawn emerged from the brush and vegetation. Dad was with me at that time.

"See them there?" Dad said. I nodded. "Now, shoot the fawn."

The gun, a large caliber bolt action rifle, was about a quarter of my own body weight. Despite my hesitation about the whole situation, dad assisted me by holding the gun by its barrel to keep it from launching me across the deer blind. So I put the scope's crosshairs on the neck of the fawn and pulled the trigger.

BOOM!

The thunderous sound of the round sent mamma deer running, but the fawn instantaneously dropped and remained motionless—a clean kill—so clean that the projectile of the bullet became lodged in its neck. The bullet didn't even mushroom as a round typically would after impacting an object.

It was a strange moment in my young life, as I would imagine it would be for most young hunters. I was ecstatic that I had done what I set out to do, but still I shed a silent tear for the deer, making sure my dad didn't see.

The only thing I questioned in my mind that day was why my dad had me shoot at the baby. Maybe it was because if I shot the mother, the fawn would end up dead anyway without the support. I doubt my dad had the foresight to think of that, so I'll go with his simple desire for tender meat.

Another thing that I looked forward to during the hunting season and being out on the hunting lease was getting to drive. Early on, I would sit in my dad's lap and steer while he controlled the pedals. As I got older and could reach the pedals, he let me behind the wheel by myself, and it was probably the most fun I'd experienced at the time.

Hunting and driving.

But hunting was still my number-one priority for many of my young years. Now that I look back on it, I wonder how much of my obsession for hunting was actually out of searching for Dad's approval.

Prior to my parents divorce I regularly hunted birds on my property, and was looking forward to the day when I could show my parents that I was capable of shooting one. There was something about stalking my prey that got my little body bursting with adrenaline. I continued trying to better my shot and then one time, while in full BDU camouflage, I walked, BB gun in hand, towards a group of six or eight birds. Crouched low, I crept up as slowly as I could. I was meticulous about approaching the group as silently as possible. I would wait a few seconds, then creep closer and closer. Finally, I felt close enough to hit the target, so I brought the gun up to fire and squeezed the trigger. THUNK! The sound of the spring mechanism sent the BB sailing through the air, only to kick up dust next to the birds. Thankfully, due to the relative silence of my gun, I could take a handful of shots before the unsuspecting birds figured out what was going on. THUNK, miss. THUNK, miss. THUNK.

A bird finally dropped! I think I started running towards the bird before the dead weight of it even had come to a complete rest. The excitement I felt in my heart was something I had never felt before.

I ran up, grabbed the bird by its feet, and ran home, which was only about fifty yards away. I was so excited to show my parents that I had finally succeeded in getting a bird, and my dad took a picture of me with one of the disposable cameras from the early nineties.

My first successful stalk.

But like I said, Dad wasn't always around since he worked as a heavy equipment mechanic at an oilfield company that his brother owned. Most days before my dad would come home from work, I would post up in the tree at the front of our trailer, where I had my tree hideout. It was the poor man's version of a treehouse, where I had a few boards to sit on, with nails placed throughout the tree to hang my guns. I would wait for my dad to drive down the 150-yard driveway, and as he drove up gravel crackling under his tires, I would jump out of the tree with toy guns in hand and pretend I was lighting him up. He'd start swerving the truck, pull it up close to the tree, and come out clutching his torso as if he were wounded. I'd run up to him but wouldn't always hug him. His work uniform smelled strongly of diesel and oil.

But though I had these whispers of happiness from my parents all throughout my childhood, I'd be lying if I told you my home life was all that great. It was a much more regular occurrence for my dad to take his anger out on us, especially my mom and sister. I was always on edge, wondering about the next fight my dad would pick.

My happiest moments in elementary were the few times my dad was a father and did things with me, because there were so few times that we actually had time together without getting in a fight. I don't ever recall him throwing a ball with me, or anything really. He wasn't interested in much, and he didn't know a lot about the things that did interest him. Still to this day, he is a micromanager and makes big deals out of small issues. Uneducated addicts with a superiority complex are the toughest people to deal with, especially if they have the means to keep their lifestyle afloat.

Shooting was the only thing my dad and I ocassionally did together.

Sometimes, there were days when I suffered mostly from neglect or carelessness. One day when I was about four, I was jumping on the bed, as most children enjoy doing. However, this time an iron was on the bed, and turned on. I'm not sure who left it there or why, but it was obviously not something I shouldn't have been exposed to as a young kid. With no care in the world for the apparent danger on the bed with me, I continued to jump. I vividly remember my mother telling me to stop because of the hot iron, and my grandmother almost immediately telling her to leave me alone so I could learn my lesson. There was no attempt on their part to get me to safety. Therefore, the inevitable happened. The hot iron tipped over onto my left foot, and I can still remember the distinct sizzle sound my skin made as it baked on the iron.

After that, I don't remember anything having blacked out from the incredible pain. I have vague memories of getting medicine from the local pharmacist and mom caring for the wounds, but to this day, some of the toes on my left foot still don't grow hair on them from the scar tissue.

The one reprieve I had through all this were my quiet nights outside. Since my childhood was during the late eighties and early nineties and I lived outside of town, there was no light pollution to distort my view of our universe. There was nothing like gazing up at the night sky and getting lost in its abundance of twinkling objects. Having a place to let my mind wander to give it a break from the weekly torment throughout my grade school years likely helped in my young development more than I can fathom.

As lovely as some evenings could be, it wasn't long before another instance of violence occurred. It usually happened when dad would come home from a night out on the town, probably partying with all the hookers, alcohol, and blow his brother could supply. Ruperto enabled my father by providing him with low-cost good times that kept him high and content with a bullshit existence. Dad could pretty much do whatever he wanted with no worry of losing his job. He worked for his brother, so partying regularly wasn't a problem. One warm night, my mother and I were asleep on the living room floor, even though our couch had a pullout bed and our window unit air conditioner was in the bedroom. I guess that she was trying to avoid what she knew would happen when he got home. Perhaps she hoped that he would take his drunk ass to the bedroom and pass out. The exact opposite happened though, and when he got home, he was his usual drunk, high, boisterous self.

"SUCK MY DICK," He boomed, looming over us.

It didn't take much to wake me up, and I remember being frozen with fear.

"SUCK MY DICK, BITCH!"

"Quiet down Rich," Mom whispered. "Rico's asleep."

"SUCK IT, BITCH!"

I laid there, hoping that I wouldn't be the next victim of his abuse. I was terrified, afraid that any minute he would explode in anger towards me and throw me around like a ragdoll. Their arguing went back and forth for a while; I don't remember much from the verbal exchange after that.

At some point, they got up, left me there, and went off into another room. I believe I remember hearing water running. Maybe she had him wash himself of any village bike he was riding.

Then somehow, the violence escalated outside. My dad had my barefoot, half-naked mom in front of his truck. The truck's engine revved and began pushing her forward, making her walk up the dirt road towards the highway. I could barely peek over the windowsill to see what was going on and saw the lights from the truck illuminating her as she cried and walked. She'd stop, and he would bump the truck into her to force her to keep walking. And each time he'd rev the truck to bump her, white dust from the caliche dirt would kick up around her illuminated body. He forced her pretty far up the driveway close to the highway, but from my limited vantage point, I could only see so much. You'd have to be deaf or dead not to hear what just transpired, and I'm sure that she finally gave in to whatever he wanted to get him to stop the abuse.

Later the truck rumbled back down the driveway. I remember being scared of what was going to happen next, so I laid back down on the ground and pretended to sleep.

But the physical torture and neglect didn't stop there. Not long before they divorced, my parents and I were coming home from doing some shopping in Corpus Christi. My sister chose not to go with us against Dad's wishes, and his anger over her presence built up the whole day. Mom and I could feel him stewing in silence. Later that evening, as we got back and pulled off the highway onto our driveway, he noticed that the lights in the trailer were on, which meant my sister was home. As he drove down the driveway, his anxiousness to get out the vehicle to put his hands on my sister was palpable.

After we parked, he bolted out of the vehicle in what seemed faster than the time it took him to shift gears, and left the driver side door open. I could hear my sister screaming gutturally, "STOP! STOP!"

Even though I wasn't there, I could imagine my sister fighting back, grabbing whatever she could and trying to hurt him with it. My mother just sat there doing nothing, and I followed suit. We both knew that if either of us tried to intervene, it could possibly be the end of someone's life. I felt so helpless; I wanted to make it stop but couldn't. My sister was thirteen, at most, when this happened. Eight years older than me.

I looked down, so afraid, torn between having to remain around the violence or the unknown of blindly running away. I knew my dad's brother, Ruperto, and their mom lived roughly eighty yards from our trailer. It seemed like a thousand yards in my mind's distressed state. But then I looked up--the field in between was covered in Texas Common sunflowers. Texas Commons grow up to eight feet, have rough leaves and stems, and take up entire areas if left unchecked. I was running the path in my young mind and pictured the worst scenarios that could happen. To me, it was like running through the Amazon Forest with a multitude of deadly creatures watching my every step, waiting for their chance to strike.

Eventually, I couldn't stand to hear the violence anymore, so I bolted out of the truck. My mom, who was still sitting in the car with me, tried to grab and stop me. The driver's side door being left open made my escape easy; I didn't want to wonder anymore.

"SON!"

I felt a slight grasp at my jacket, but I got away. I ran through the field, doing my best to shield my face by holding my forearms together in front of it. I ran as fast as I could because I was afraid that if I didn't, a rattlesnake would strike me as a slow-moving target. I could feel the jagged edges of the rough sunflowers scratching away at my fragile skin, and the only place I slowed down was when I thought I was getting closer to the horse track. This was the first time in my life that I felt the intensity of adrenaline push away my fear. I somehow knew that as long as I kept my heart beating fast, my dad couldn't catch me, and couldn't beat me senseless.

The metal pipe that was the railing boundary for the track was head-level for me, and the sunflowers grew above, concealing it, so I needed to make sure not to clothesline myself. Once I got through to the track, it was an easy run to my grandma's trailer, where I started banging on her door, begging her to let me in.

She eventually let me in. I was there with her crying, and a few minutes later, my dad showed up with his face damaged and lip busted. I'll never forget the look in his eyes. They were the eyes of someone at war with the drugs in his body. He wanted me to go, and I didn't want to. I wanted to stay with my grandma.

Eventually, though, I gave in. After all, I was always a daddy's boy.

CHAPTER TWO

DIVORCE DAYS

As you might imagine, Mom eventually had enough of my dad. I didn't see much of what happened, but I was spending all my time on the ranch before I knew it.

They separated not long before a few incidents involving my dad pointing a gun—vaguely or not—at my mom whenever they crossed paths. Normally I tried to keep my mouth shut, but once they were divorced, I let my dad have it when I was living with him in around third grade.

The major incident occurred later in the evening, when I was riding in the middle of some sort of single cab or extended cab truck. Some guy was driving: whoever my mom's squeeze was at the time, and my mom was in the front passenger seat. I think we were driving from Corpus, and she was taking me back to my dad. I remember us pulling up to the stairs leading up to the door of our trailer; the wooden deck my dad had built a few years prior was gone, and in its place was a generic metal staircase you might find at an oilfield site.

Any vehicle pulling onto our property easily drowned out the single incandescent light on the trailer, and dust from driving down the dirt road kicked up everywhere. Dad and I used the back door of the trailer as the main door, so a hallway was what you immediately walked into with the master bedroom and only bathroom to the right. The other bedroom, living room, and kitchen were to the left.

Before we could make it out of the car, the trailer door swung open. There was my dad, preparing his usual addict rage fit. He started to yell and curse at mom as she got out of the vehicle. I hopped out, too, with my head down, not wishing to see anything. I passed by my dad and beelined it straight for the primary bedroom, hoping that the violence wouldn't escalate.

Within what seemed like a few seconds of being in the bedroom, BANG!

My heart stopped, and my feet somehow managed to carry me over to the doorway, looking into the hallway where my dad was standing.

BANG! The second round went off—amplified by the tin can we lived in—and my ears rang with a high-pitched sound. I saw my dad with the revolver in hand, and his body language was an image that I'll never forget. It almost seemed like his entire body was pulsating, probably a side effect of endorphins from the drugs coursing through his body.

Shortly after the second shot, I heard the vehicle spin its tires as the driver quickly got the hell out of there, for obvious reasons. After he backed out and left, my dad slammed the door and went to our kitchen table to continue drinking and probably get another bump in. There was so much running through my mind; I didn't know how to process it all. He had been preaching to me about gun safety my whole life—I mean he DRILLED that shit into me. So much so that even when I snuck away to admire the bolt action gun in my dad's hiding spots, I never touched it without him around. Due to his paranoia of retaliation from his "enemies," he kept his guns in easily accessible places, so often I would go and take a peek. But when he shot at my mom, I was pissed. I went over to him while he was sitting at the table, and I gave him a piece of my mind.

"What are you doing?! Why did you shoot at mom?! You're not supposed to do that! You're not supposed to point them at people!"

I let him have it, and his response was, "It's okay, I just shot at the radiator."

That was it; I didn't know what else to say after that. He could've said a million things, but he straight up told me what he did with no remorse. I probably didn't even know what a radiator was at the time so that just further dumbfounded me. I remember turning my back towards him and walking away in anger. I was so tired of these people acting like children, and I was a child! I'm fortunate that the man my mother was with at the time didn't have a pistol and retaliate in fear for his life, which is something that happens more often than not today.

This incident was the first time I was forced to deal with altercations between my bickering parents. Before, I was able to avoid it most of the time. It was always something with those two, always acting out when in each other's presence, though my dad took all responsibility for the violence. Mom just attempted to defend herself with what she had, which wasn't much.

Another time I had to deal with this was when Mom dropped me off at Dad's, but instead of shooting at her, he went back to common assault. They got into an altercation where he bit her finger, pushed her down the steps, and who knows what else.

I hadn't even gotten out of the vehicle, or if I did, I got back in shortly after the fighting started. In the truck, we left the ranch and pulled over. We were on the road headed West towards Freer. I believe they waved down a Texas State Trooper because mobile phones were only for the one percent in those days. One eventually stopped by to take a statement from my mom and get pictures of her finger where Dad had mangled it from biting her. He flashed his flashlight over the bloodied finger; to me it looked almost pinkish, like meat. After some time with my mom, the officer turned his attention to me.

"Son, do you want to go home with your mom or dad?"

A flood of emotions came over me. I didn't know what to say. Just the fact that the police officer would give me a choice to go back with my dad, who had just committed an act of violence, made things that much more confusing. I was such a daddy's boy but knew what he did was wrong, like most things he did.

Even then, though, I didn't want Dad to be home alone because I didn't want him to do something crazy like killing himself. After all, his son chose not to be with him. It never occurred to me that he probably didn't expect or even care if I came back with him. I would've just ended up alone in my room playing with video games or Legos while he drowned himself in narcotic substances, like usual. I didn't want to pick him and hurt my mom—it was beyond me how she even dealt with the abuse and then watched her son go back with that monster without saying a word. I didn't know what to do, but I knew I didn't have much time to decide. I felt like I was being torn in half.

But being the mature man I knew I had to be, I chose to go home with my mother that night. But I still feared for my dad's safety.

I told the officer that I wanted to go with my mom. I fell asleep in the car on the way home, leaving Mom to sit in silence.

*****

Somehow, my mom knew exactly when and where Dad was cheating on her and one day, she decided to intervene. She dragged my sister and me up from CCD and had one of her friends drive us to Corpus Christi to catch Dad in the act. I knew something wasn't right, so that ride was another anxiety-filled trip. I remember us stopping at a gas station, and then the seemingly long ride to my grandparents house in the "hood." I didn't see exactly what happened, but as you might guess, things got ugly. They fought all the way to the car, and he joined us on the drive to my mom's parents' house, which was in Corpus. The whole way, of course, there was screaming, crying, and fighting. As my father drove recklessly down the freeway, I stood in between my mother's legs holding onto the dash, crying frantically with her in the passenger seat. I remember hearing my mom tell him,

"You see what you're doing to your son and family, Rich? Do you see?"

She repeated that for the entire short trip. He kicked us out of the vehicle when we got there, and I remember him driving off down the road and my sister chasing him, beating the vehicle, and just cursing angrily. She was ready to kick the car door in and dent it. After the altercation, we went inside, and from that day on, my grandparents' home was my new place of residence.

Back at the ranch, my dad had been building a house right next to his trailer to finally move into something that wasn't going to rock with every bad storm. I remember staying in it a few nights before this entire mess, but it was just a shell; we still used the trailer for our cleaning and cooking. He worked on that house occasionally during his time off after work, and after she caught him cheating, my mom went and tried to burn it down. She dragged my sister along with her, for whatever reason. I remember staying in it a few nights before this entire mess, but it was just a shell; we still used the trailer for our cleaning and cooking.

My sister told me about it later. Mom had grabbed a gas tank from out of the back and ran inside to grab the matches. But after setting the first little flame, my sister finally noticed what was going on. She helped put the fire out, so it didn't burn to the ground, and my dad eventually gave the house to my cousin.

After that, Dad was back in the trailer and we stayed with our grandparents in Corpus. Dad was a ninety-minute drive away, my sister was in junior high doing her rebellious thing, and my mother went to work. From that point on, I was a mess. I cried all the time, and the only person who was there for me to cry to was my sister, which was only after we initially moved. I was essentially alone. No one was around to see if I was doing okay, so I had to take care of that myself.

There were many people—mostly family—around in that house, approximately fourteen people at a time. It was a tight squeeze, to say the least, in a 1500-square-foot house. But despite the little privacy I had, this was a time I felt most invisible since everyone likely assumed that someone else was taking care of me. The truth was, no one except me was doing that job.

My time living with my grandparents was relatively short, but the things I went through and saw in that short time showed me that adults in general, not just my dad, have more issues than Playboy magazine. While in Corpus, my mother and I lived with one of my aunts and her son for a bit. I had no idea where my sister was staying during our stay there. At the time, though I didn't know the woman we were staying with was my aunt, mom never communicated adequately to me. This made me feel that much more awkward when my mother left me there alone with those who were essentially strangers, even if they were family, since I didn't know them. I'll never forget the hurt I'd feel being alone at night staring out the window from bed in a strange place, hoping that every car that drove by would slow down and pull into the driveway.

CHAPTER THREE

GROWING PAINS

Not long after this, my junior high years began. I left my grandparents' jam-packed house, spent years alone through elementary school, and moved with mom and sister into another single-wide trailer on a small lot towards the center of our small town in 6th grade. The years I spent in town were some of the best in my life and the worst. I wasn't stuck on the ranch anymore and could easily walk to many of my friends' houses within thirty minutes.

After the divorce, my mom worked three jobs to provide for my sister and me. She worked at a convenience store and as a bus driver and cafeteria lady, all with little support from dad. When she wasn't working, though, Mom was usually with her boyfriend Raul, and they were usually either at his ranch, in Mexico, or out and about spending money somewhere. He was a well-off man, likely from the oilfield, so she was usually out with him.

One of the best things that happened to me that year was when my mom bought raffle tickets from the local Border Patrol station. The grand prize was a computer, and she won! Finally, we had a PC to experience the first days of dial-up in my town. The only time I got with a computer came during school hours before that. There wasn't much I could do on the internet in those days because the dial-up was so slow, most of my time spent online was playing games or in chat rooms. One of the first things I did was teach myself to type without looking at the keyboard. I was addicted to busting my ass to try and be the fastest in the class. I was pretty damn good, but one person was better: Danielle. I typed around fifty words per minute, but Danielle could type in the low- to mid-sixties. I never beat her but was glad that I could hang with her.

The single-wide trailer we lived in was just a place for my mom to call her own, and the prominent residents of it were my sister and me. It was not too different from the one we had lived in prior, except it had two bathrooms instead of one.

My whole young life, I never had my own room. We had always lived in two-bedroom trailers: one bedroom reserved for the parents and the other for the oldest sibling. I slept with my parents' until my sister moved out, and it took a while to want to sleep alone. Eventually, my mother bitched at me to go and sleep in there. As scared as I was, it didn't take long to find comfort in sleeping in my room. I even slept in there during the day sometimes; looking back on it, I slept a lot of my young life away. I wasn't studying; no one expected me to, so why would I? I would much rather shut things out by sleeping or messing around with video games.

My mom always wanted me to be home before dark, but as long as there was light in the sky, it wasn't dark to me. Whether I was with friends or by myself, I enjoyed the thrill of a few moments of freedom, especially since I knew that breaking the rules wouldn't mean getting slapped or violently yelled at. Mom would occasionally get mad at me early on about staying out too late, but she was at least fair about it. And as I got older, she trusted me more and became more comfortable with me being home alone.

I excelled in junior high with no sincere effort and no support system. We took a computer test to gauge our reading level at the beginning of the year for reading class. As a sixth-grader, I tested at a senior reading level. When I sat down for that test, I thought about purposely doing wrong on it to make my life easier, but I couldn't bring myself to fail intentionally.

I thought my reading skills were impressive at first, but when I couldn't read books with pictures, things quickly turned into a nightmare. I started reading the lowest level of books and did the minimum number of books. I read a lot of *Hank the Cowdog*, a series of books about a dog who worked the ranch and had adventures along the way. It had a few pictures, but after reading most of that series, I went up in book level, and with that went the pictures. I don't remember reading many other books, but the best one I read that year was *And then There Were None* by Agatha Christie. I'd take the book quizzes and still pass them without thoroughly reading the books. Another time in that same class, I walked into quite the surprise. We had taken a test or quiz a few days prior, and I had scored the highest out of all the classes, including the advanced placement students. The teacher wrote big on the chalkboard that I had scored the highest for all the students to see.

I don't know why she did that. Maybe to show everyone that if I could do it, so could everyone else. But more than likely, it was to try and make me feel special for a day. I'm not a trained educator, but I assume that it doesn't take long for teachers to know who the less privileged kids in the class are.

I couldn't believe it when I walked into the classroom to think I had beaten the best of the best: Angelina and Danielle. Salutatorian and valedictorian, respectively, were always top of our class.

But as good things come, so do bad things. The worst part of class was when my cousin would embarrass me by asking why my canines were so high.

"Aww, I can see your canines peeking through your lips."

From that point on, I made a point to try and not smile, as ridiculous as that sounds. I could never smile with my teeth, and I was always conscious about it because I didn't want the embarrassment of having to answer questions about my teeth. As though I had some control over it! I always figured that they would eventually come in correctly, so I responded similarly.

Come to think of it, I had never been to a dentist and had no dental hygiene habits. Maybe I brushed my teeth at night, but no one ensured that it was getting done during most of my childhood. Things changed, though, when one of my teeth fell out. Upon inspection of the tooth, I noticed a cavity had eaten through it, basically hollowing it out. That hollow tooth scared the life out of me. As with most things, I didn't tell anyone, and from that point on, I ensured that I brushed my teeth every morning and night because I didn't want all my teeth to be full of air.

There were many crazy things we did as junior high kids: broke into abandoned homes, garages, buildings, and stole small stuff from convenience stores occasionally.

However, the most reckless thing we did was shoot our BB guns at .22 caliber rounds in the hopes of setting them off. We only did that once, though (successfully, I might add), since we didn't want something severe to happen. We weren't even sure we could make it happen at all as we were shooting from a distance and behind cover for our safety. Zac was one of my best friends in junior high and was a regular participant in these activities. He lived about one hundred yards away from me, right down the road from Kurt. I also often walked or biked across town to Julian and Conner's house to hang out with them.

So as you might expect, in junior high, I missed a lot of school, most of it for no real reason. My mom's go-to excuse for the unexcused days was to write this note:

"Sorry my son missed school, but his [insert random relative here] passed, and we were at their funeral."

As an eighth-grader getting ready for freshman year, I told her to stop because things might start getting obvious, and it was ridiculous. It wasn't the right thing to do in any case, but she just brushed me off like it was no big deal. I missed so much school in my sixth-grade year that I had to do summer school. It was the most annoying thing ever; all the work was so easy that I basically breezed through it and just sat at my desk bored.

Being home alone at night became a regular occurrence—and sometimes those nights seemed surreal. My sister had moved out of the trailer by this point, and her room became my room. The only interruption from the quiet was the intermittent rumble of cars on gravel outside; the trailer also had some weird sounds: creaking, plinking. Even little noises like that kept my eyes wide open, and I felt anxious for no reason at all.

One night, in particular, I walked from my friend Kurt's house after hours of playing a popular zombie video game. This video game series was a staple in my life, so its influence played heavily in the sounds I heard that night or my interpretation of them at least. After being freaked out for hours, I didn't want to go back to my trailer. The darkness outside made me genuinely scared that I would be attacked. Despite my fears, though, I went home. I can't remember if I walked or ran the twenty yards back; I probably ran.

The night wore on. I had school the next day and decided to sleep in my mom's room. I was so scared that night I wanted to fall asleep with the TV on, even though it was something I never did. It might've been the fall season, as the windows were open that cool night to save money on air conditioning. Laying in my mom's bed and halfway watching *Nick at Nite*, I got sleepier and sleepier but fought it.

As I forced myself to stay awake, it got to the point where I started to hallucinate and hear things. It sounded like someone was dragging a heavy plastic bag or something across the backyard. Scared out of my mind, thinking there may be something supernatural in my backyard, I pulled the covers over my head and forced my eyes closed, trying to fall asleep so I wouldn't have to think about the fear.

Suddenly, after who knows how long, I woke up because I couldn't breathe. I had my rope chain necklace on, which I usually didn't sleep with, so I sat up thinking I was choking on my necklace. When I sat up, I was still unable to breathe. I began grabbing at my necklace with no success at getting my fingers underneath whatever was around my neck.

I tell people that Duval County is the most lawless place I've ever seen other than Afghanistan, and that statement holds to this day. I'm sure there are other small backwoods places in the middle of nowhere that have just as much corruption, but this is the only one I've genuinely experienced.

Okay, what the fuck is happening, I thought, as calmly as I could.

I tried to turn my neck to see what was going on, but mind you, I was still 95 percent asleep and had no fucking clue what was going on. As I tried to turn my neck, there was so much resistance around it that I couldn't move.

The light from the TV began to fade, and I knew I was running out of time. As a last-ditch effort, I remember thinking, "I don't think my necklace is hung on anything." With my left hand grabbing at my neck, trying to get under whatever was choking me, I stuck my right arm straight out, and with all the force I could muster in my little body, threw my right elbow directly behind me. I remember making contact as my elbow didn't freely fly behind me. The choking finally stopped, and I heard a shuffle as the perpetrator quickly ran out of my house. As I sat up in bed, I was still unsure of everything that had just happened. I sat there for a few minutes, trying to analyze what just occurred, thinking maybe it was a dream. Shortly after, I went to the kitchen, grabbed the first knife I could find to put under the pillow next to me, and went back to bed.

The morning was calm as ever. I woke up and suddenly remembered what had happened the night before. I looked around to confirm that what happened last night wasn't a dream. I walked from my mom's bedroom through the house and didn't see any signs of a break-in.

It wasn't until I got to my room that I found where they got in.

They came in through my window and knocked over my bookcase. I went outside on our deck and checked to see if the key we typically left outside for each other was in our hiding spot. It was gone, so I decided to go and inspect the damage I sustained.

In the bathroom mirror, I could see I had bad rope burn all around my neck. I guess this was from trying to turn my head to see what was going on.

When my mom got home, I informed her of what happened, and she just changed the locks to the house. I don't think she ever made a police report because I think she knew that nothing would come of it.

The scarring on my neck was pretty bad. It scabbed up, and after the scabbing healed, the scars were still visible until late into my high school years. After that, I was pretty scared to stay at home by myself, obviously, and my mom tried to accommodate staying at home with me, but it was just a matter of time before I was home alone again. No crimes ever happened again, but I usually found myself awake most of those nights, sweating because I was paranoid about forgotten unlocked doors and windows.

Because there were no other options, I assumed I just had to "get over" any fear of staying home alone. If my dad were still around, he'd probably tell me to suck it up and deal with my fears head-on. I tried that, but nothing helped me get much sleep at night.

I kept trying to stave off the anxiety as I entered eighth grade. It was the last year of Junior High, and while I didn't play football that year, I replaced that physical activity with a new video console. That year I switched from video game consoles after an even trade with my friend Martin. Even after the break-in, I stayed a true gamer and fanboy for life by playing horror games late at night, alone at home.

Video games have always been an essential outlet for me, especially with the introduction of online gaming in the early 2000s. I remember my mother buying me one of the newest video game consoles from a guy we knew, so she got a bundle with a few games, and it was great having the most powerful console at the time to play. I spent so many hours playing the great games that the console had. I even bought the most popular shooting game of the time before I had the console because Conner's cousin had one, so we were over at his place all the time playing it.

I spent hours playing these games, which kept my mom happy because if I was in my room playing games, that meant I wasn't out on the street. Getting into fights wasn't something that I mainly wanted to participate in, but it occasionally happened.

One of the more reckless things I did around this time was join in on BB and paintball gun wars without any safety equipment (not that there is any for shooting BBs at each other). I'll never forget when I had my one-pump BB pistol that was pretty weak, and my friend Kurt was on the roof of my backyard shed without his shirt off. I was a fair distance away from Kurt with my friend Chris and he had the BB gun.

"I'm gonna take a shot at Kurt," he exclaimed.

Twitching his wrist from his hip, pistol in hand, he fired into Kurt's general direction. I saw the BB flying through the air away from him when all of a sudden, it seemed to take a 90-degree turn! The Magnus effect took over and curved the BB significantly. The Magnus effect, in general, happens to spinning objects in a fluid. Since BB's are round, they regularly encounter this phenomenon and take curved paths over long distances.

"Ah, you *motherfucker*!" Kurt screamed. "You hit me on my nipple!"

"No way," I thought to myself, and he hopped down to show us. His nipple was extraordinarily huge, confirming the direct hit. It blew my mind, mainly because I saw it all happen.

After that incident, I don't remember playing much BB gun war; I probably subconsciously stopped after seeing how much fate factored into getting hit with a BB. Two more significant BB gun incidents did happen, however.

The first was when one of my classmates was shot with my BB gun in my home by two older little thugsters. My friends never hung out with these guys because we knew they were trouble. Many people were over at my house, and those two showed up and came in while I was busy doing something; I think I had run over to the main restroom for a quick pit stop. When I came out, there the troublemaker was, pumping my BB gun and shooting it at the calves of one of my classmates. I was sure it had no BBs in it, and as he shot my classmate, he was writhing in pain. Again, he started pumping the gun.

Pump, pump.

Thunk!

Pump, pump.

Thunk!

The victim didn't seem like he was serious; I thought it was an act the whole time. But the thing is, the golden rule of gun safety is to treat them all as if they are loaded, and it wasn't until about the fourth shot that the victim finally yelled,

"Cut it out dude!" The troublemaker was preparing the gun to fire again.

At that point, I realized it was no act and went to intervene.

As I approached the perpetrator, he handed me the gun, and the two ran off. It wasn't until my friend lifted his pant leg that reality sunk in. He was bleeding from a few different spots on his calves. He immediately went home, only a few blocks down the road. Shortly after, like an hour later, at most, the police came by my house and confiscated my pellet gun. I was pretty upset about the whole thing, to begin with, and then to get my gun taken away—which was both pump and CO2 operated—was aggravating.

The entire time I had thought that kid was kidding and that the two of them were trying to trick me or something; I didn't know he was actually under attack. Especially not after the first one because the guy getting shot could've easily defended himself. He was one of the bigger guys in our class.

My close call with a BB gun happened in my friend Kurt's bedroom not long after the BB-in-the-nipple incident. I was over at his place pretty often. One time as we walked into his room, he decided that he would point his BB gun directly under my chin. Faster than he could put it under my chin, I swiped it away.

"Don't point that shit at me," I said. "You don't know if it's loaded!"

"No it ain't, look."

He pointed the gun at his palm.

"THUNK!"

The gun went off, and he immediately dropped it, grabbed his wrist, and started screaming in pain. I was both happy and sad about what had just happened. I ended up without a BB in my neck, but my friend, to whom I was trying to teach a lesson, had ended up teaching himself the hard way.

I have always been about gun safety, and when I participated in the BB gun wars, I never directly shot at anyone. I knew the dangers of what we were doing, so I'd stay behind cover 99 percent of the time, other than when I'd pop out to pretend to be shooting at someone.

Paintball wars, on the other hand, were a different story. We played those without protective gear as well, and I spent a few days setting up the backyard behind my single-wide trailer to be a mini paintball area. I had dug trenches to put whatever scrap I could find to make a cover for us, along with pallets and other supplies. It was a pretty good time, though it sucked when we got hit. Thankfully, no one got seriously hurt while paintballing at my house, and the worst shot I ever received was on soft tissue near my kidneys.

The unfortunate thing is that there were more severe injuries that occurred to others while playing these games, but it wasn't long before the town's luck ran out. Some happened by accident, not while playing games, but they're all very tragic regardless of how they occur. It happened to children much younger than me, but nevertheless, the access we had to BB, airsoft, and paintball guns are somewhat frightening now.

CHAPTER FOUR

GETTING WIRED AND DIAL UP

Time went on in a bubble of swirls and panic, so I sought out friends as much as I could to get away. Mostly, I looked for friends who had video games we could play since I was sure as hell not going to sit around and tell my friends about my problems.

My friend Jeffrey had an ambitious console pushing graphics to the next level and was an excellent console for its time; we played it quite a bit. We played basketball a lot, too; it was a big part of the community for the young men to be out playing. We just casually played outside, no matter our age. We often played street football too—the way we played was "two-hand tag" to down the player. People of all ages and capabilities would get in on it, and it was a great time. However, I did have my disagreements with the other players, and we would sometimes end up mad at each other. Unfortunately, living with my mom didn't keep me from domestic violence, as my sister's boyfriend beat her regularly. I recall calling the police multiple times after he battered and bruised her badly.

I remember the most prominent time when my sister pulled into our parking area. They got out of the car and circled it because her boyfriend had tried to beat her. She cried out to me to call the police, so I did, and a few minutes later, they showed up to intervene. I remember wishing I was bigger so I wouldn't have to worry about calling the police; I would go out there and take care of him myself. Trouble was, I was so small up to high school; I only weighed one hundred pounds at the time, so doing physical harm to someone a decade older than me was outside the realm of reality. My sister and that guy were always getting drunk and doing dumb things, so him beating her was just part of their routine. How my mom allowed him to be around—let alone live with us for a bit—is beyond my comprehension.

Looking back, it makes me think that I must have been way out of touch with my emotions since I couldn't tell most of the time if I felt sad, angry, or both. Eventually, sadness won. There were times when a friend's words would cut me to my core, and I'd instigate a fight. I remember once passing by Jose's house and the extra property they owned—a fenced-off lot with a wooden playground with slides and a swing in one corner. There was a shed for his father's stuff in another corner. It was a reasonably big lot—I'd say a little over an acre—so big that we'd usually play on the lot when we played baseball.

I don't remember what happened or how we got to this point; I would guess that they lied to me about something Jose's little brother said about me or "my mom." He was playing at the wooden playhouse, and I was at the outside of the fence, hollering at him to come out so we could fight. Though he was younger than me, he was still bigger. I weighed ninety to one hundred pounds throughout junior high, so there weren't many boys that I was bigger than, regardless of grade. Despite this, I still wanted a fight.

"Come here!" I shouted. "I just wanna talk!" I knew that was a complete lie, and so did he.

That's when a kid named Gaz—an acquaintance I didn't know too well—came walking down the back alley behind the lot. Gaz offered to take his spot, so I complied; my mind set on fighting someone, anyone.

This was a huge mistake. Gaz had younger and older brothers and likely had thousands of times more experience fighting than I did. All I remember was him jumping the fence from the other side, walking over, and then being in front of me. It was all I could do to drop my head and throw some punches, but I couldn't hit him. He could so quickly move his head off the centerline, then counterpunch me in the face since I was standing square in front of him, feet cemented to the ground. I was so confused. He wasn't walloping me, but I couldn't hit him, and he was jabbing me little by little every time.

I don't remember how many punches I threw or how long this went on, but eventually, I plopped on my ass and started crying out of frustration. Thankfully Gaz didn't do any severe damage. After we separated, Gaz ran off, and I went to Kurt's house to recover, which didn't take long. I sat around until I could catch my breath from crying and the fight. I was in no physical pain. All I did was wallow in humiliation and confusion.

The worst part was that Jose's little brother, the one I was initially trying to fight, witnessed the whole ordeal. After that, he thought he could take me.

A few days later, we were at Kurt's house. Jose's brother walked straight up to me and pushed me, hoping for a swing back at him. Once again, I obliged.

Kurt and our other friend Zac had previously coached me on what to do and how to fight, so I employed what they taught me. I went in, grabbed his shirt collar with my left hand, and started punching him in the face with my right. We were standing for a minute with me landing a few shots, and then ended up on the ground. I had him mounted and started punching him in the face again when I was suddenly pulled off by Kurt and Zac.

What I did to Jose's brother didn't feel very good, but it did feel good to know that I was getting better at defending myself. He likely just felt like I did when I fought Gaz. I knew the physical damage I did wouldn't last long; I had the strength of a wet noodle.

I couldn't avoid fights, though, and I had more in junior high. They seemed to follow me, even though fighting wasn't something I was good at or even knew how to do. My go-to move for most of my life was to stick my arms out, tuck my head, and charge until we ended up on the ground. Unfortunately, as the years passed, I found myself in more fights, but on that day, when we weren't playing any games, I uncharacteristically tried to be a bully and ended up being the one sent home crying. Though I didn't know it at the time, every altercation I was getting in was critical in preparing me for the attacker at home.

Something about fighting seemed to scare the competition out of me, so I dreaded going into the seventh grade because that was the age athletics started. My dad never played catch with me, let alone teach me anything about any sports. To this day, that old geezer still doesn't know most of the rules to football, a game he has watched most of his life. He can tell you how many quarters there are, how many yards to a first down, or how many downs you can take before a turnover, but he likely couldn't tell you anything "deep" about the game, like penalties or playoff seeding.

I took after my dad in this way. And as I entered seventh grade, things turned out to be worse than I expected because I expected the coaches to coach us. Instead, it mostly seemed like they just had us hitting each other, teaching us about the routes and positions, but nothing about rules. I didn't care to know any of the rules or play anything.

I couldn't distinguish the rules from improper coaching during my first time playing. As with most small players with no crazy speed, they delegated me to be a receiver, and I lined up anxious for my first play. That was when the whistle blew. My mouthpiece wasn't in, so they threw the flag for not having it in my mouth. I couldn't believe I forgot to put it in, but it wasn't too surprising, seeing how I wasn't used to having it in the first place. I got pulled out of the game after that, and the first thing our coach told me was,

"If you do that again, you'll never touch the playfield."

I was taken aback by his demeanor, especially since "coach" never mentioned that not having your mouthpiece in was a penalty. I didn't need another instance like that never to touch the playing field again; the coach just never bothered with me or my learning after that first incident.

Things got to the point where I realized that he would never play me, so going to practice and putting up with the shit I had to do was not worth it for me. I remember one morning outside the cafeteria in my mom's maroon single cab truck, her maroon polo uniform crisp and ready for work. I was hunched over in the seat crying, wanting out of football, insisting that I didn't want to participate. She did her best to console me, but she never took steps to get me off the team, and I had to endure the rest of the season.

I always preferred to spend one-on-one time with my closest friends. I spent a lot of time in junior high with my friend Jeffrey. He lived a bit further away than most of my other friends, but I went over a lot more often once I got a bike. His little brother was a pain, always punching people in the dick. He would get his ass beat as a result, but you would think he would've learned to stop hitting dicks to stop getting beat.

He got me good a few times, but I usually did an excellent job protecting myself. The dick-beatings were so bad throughout elementary that I remember some nights I slept covering my balls with my hands. When visiting with my Aunt and sleeping over at the neighbor's house, people ask why I slept holding myself. I told them it was muscle memory from having to protect myself at all times in fear of someone coming up from behind me and "racking" me.

And things kept changing. On the way home from school—not long after getting the internet—I found a small bag of cocaine in our driveway.

Hey! Is this what I think it is? I asked myself. *It would be a great idea to test it the way police do on TV.*

I saved it for trial, and when that later date came, I made sure to have a friend there to witness it. I remember being in front of the computer, licking my finger, and then dunking it in the baggie. I went wrong by licking too much of my finger, which put a lot of powder on it. I licked my finger without a second thought, and my tongue immediately went numb. I had no idea what was going on, and it was a bizarre feeling. I bolted for the refrigerator and grabbed the chocolate syrup we had in there, and just started pouring it into my mouth. I was bouncing off the walls for the rest of the night, high on coke.

CHAPTER FIVE

A LITERAL FIGHT FOR MY LIFE

My romantic life at this point was pretty much nonexistent. Without the confidence to smile or even sometimes look at others, I relegated myself to simply not concern myself with such things, even though I wanted companionship as much as most people do. In my mind, I'd always been pretty much the bottom of the barrel, worthy of no one and barely worthy of breathing a more capable person's air. At this point in my life, I remember telling my mom,

"What's the point of doing good in school and learning? We're all going to die anyway."

She spoke about working hard so I wouldn't suffer and trying for a better life. But her advice didn't do much good at the time—I felt like no one would ever want me and that everything was pointless. There were some girls I knew, though, whose names I'll never forget.

Angelina had a crush on me in seventh grade, and I had one on her too, but I knew that she wouldn't stick around long if she got to know me. So once I knew of her interest, I kept our relationship as platonic as possible. There was another girl a bit younger than us who also had a crush on me, and I ended up "dating" her for a bit, though it was nothing more than a handful of times spent together at the little league park. After word got out that she and I were boyfriend and girlfriend, I received a phone call from Angelina. Of course, she was unhappy with me, and I felt terrible for many reasons, but mostly because I wanted to be with her, not who I ultimately got with.

These complicated feelings all changed when I saw Celina for the first time. I remember getting the "I know you from somewhere" feeling when I first laid eyes on her. That's when I thought I was in love for the first time. Upon seeing her, all my insecurities fell away, probably because she was in sixth grade and I was in the eighth. In reality, it was mostly because her smile and demeanor. She was shy like me, especially when I began flirting with her in the halls by saying, "HI Celina!" and waving at her. I'm sure it was embarrassing for her, looking back on it. I wanted her more than anything else I had ever wanted in my few years of life, and I was going to flirt with her as much as I could to try and be her boyfriend one day.

I tried everything that entire year, including getting my friends to give her my gold rope necklace. At the time, that necklace was one of the only things I owned that was worth something. Not surprisingly, she declined the offer. I'd call for her attention; she'd look away. She wanted nothing to do with me, but for some reason, with her, it didn't bother me. I don't know how or why, perhaps because I felt more mature and she was younger. Celina was the first person who gave me the confidence to act like a fool. Though I was interested in a handful of girls, and there was a handful who liked me, I would've done anything to be Celina's boyfriend so I could just talk to her.

That never happened. Once I finally realized I could never have a chance with her, I decided to try things with Angelina. We ended up "official" just before our freshman year. At our eighth grade dance, I danced a lot with Angelina even though I didn't know how and didn't even want to. I just wanted her to know that I did like her a lot.

In the end, though, I gave up on most relationships simply because I didn't have the tools to be a proper boyfriend. I barely had the tools to operate on a day-to-day basis. This was especially apparent to me after getting into another fight, which resulted in a severe undiagnosed injury.

It happened one afternoon after school when a few of us gathered in front of the band hall. For some reason, my other friend Donald and I had begun to fight, and he outweighed me by forty to fifty pounds easily. I somehow managed to get him on his back, and I was standing over him. Still, Donald overpowered me: he held me by the wrists with ease, since my wrists at the time were probably a half-inch in diameter, so he easily kept control of my arms and used his feet to topple me over.

With that, all ninety pounds of me began the headfirst journey to the earth without my hands to brace my fall. I remember the world turning upside down, and—closing my eyes the last second before impact—everything going black.

We were on the caliche portion next to it, but it was still rock hard. Luckily we weren't fighting on the concrete. I have hazy memories of dragging myself up after that and walking away towards my friends' house since they lived within 150 yards of the school property.

With Donald's last blow in that fight, I received a severe concussion that went undiagnosed. Even the most minor bright light or sound brought on an incredible migraine; the pain was unbearable. I spent a lot of time at school and home in horrific pain and would plead for my mom to please take me to a doctor. I don't remember how long the headaches went on or how much I had to beg and plead before she gave in.

After what seemed like at least a month of agony, we took a forty-five-minute drive to the pediatric clinic in Alice to see a doctor. I was finally relieved to see a medical professional and expected to get help, but I'll never forget what he told my mother and me. I don't recall if he asked me about any significant hits to the head (I don't think he did), but I explained all my migraine-like symptoms. His response?

"Only females get migraines, so whatever you have is something, but not a migraine."

The doctor sent us packing with the wrong information and no actual diagnosis or prescription. Of course, my mom wouldn't question what he said and would not take me elsewhere for a second opinion, so I continued in agony. The migraines went on for weeks after that, but it was rough going for a long time while concussed, especially since video games were my outlet. After playing for a bit, the light and noise sensitivity got to me and left me confined to my bed with something covering my eyes. Those were rough days.

And then, one day, they stopped. I don't know when I realized that I stopped getting them, but it was apparent that they were gone. To this day, I haven't had a headache or true migraine like the ones I had then, as though I lived through a lifetime's worth of head pain in just a few months.

CHAPTER SIX

BACK TO SOLITUDE

Freshman year, I lived with Dad since Mom had sold her property and moved to Alice with a new boyfriend. He had a few acres outside of town, and they were planning on building a house out there. At the time, they lived in a larger-sized single-wide trailer. My mom pleaded with me a lot in my eighth-grade year to move with her to Alice so that way I could have a better quality of life as opposed to living with my father, who was still a partying addict. I refused; there was no way I could leave my friends, who were essentially my family. I wanted to, but being an introvert, I didn't want to be the 'new' guy at school and deal with all of that.

I stayed, and living with my dad afforded me the freedom to do whatever I wanted. At some point, before I moved back in, my dad had bought a new double-wide trailer. It was a three-bedroom, two-bath, and it was drastically bigger and nicer than anything I had lived in before. The single-wide we previously lived in was taken off the property, and all that was left was a laundry shed that he had made where we did our laundry because there was no room for it in the single-wide.

Mom's boyfriend, whom she would eventually marry, worked for Halliburton, an older man who had been working there for many years. He had quite a few kids; I'd say anywhere between five and seven of them.

Over the summer before freshman year, I spent most of my time with my mom in Alice and didn't do much except help Mom and her boyfriend work on their house, play video games, and eat a lot. At that time, Mom cooked a lot because she didn't have to work. I never before had access to this much food! I ate so much it got to the point where Mom told me to suck my thumb because she wasn't going to cook for me right then and there. I ended up gaining around twenty-five pounds that summer, which brought me up to a whopping 125 on a good day. So I went into freshman year with a little meat on my bones—not much, but it was better than the hundred pounds I weighed before that.

But the summer of leisure didn't last, and I realized that high school had arrived before I even noticed. I met the girl I dated for most of my high school years early on at the football games. Her name was Monica. She was two years younger than me and had two little brothers who were a lot younger: one in elementary and the youngest not even in school yet. After meeting Monica's mom in public at school or the games, they invited me over to go and hang out at her house.

Freshman year, I decided that I would play one sport a year, and that year it was going to be baseball, as I was still pretty small for my age. I revered the kids who played every sport every year since I had enough trouble getting to practice for just one. Not only that, it was crazy to me how the other kids could start practicing for another sport before the current sports season even finished.

Transitioning into the academic-athletic world was something way more difficult. The only thing I had learned about sports was on the streets in junior high during this time. I worked out with the team for a few weeks but never got to play because of failing classes. I failed classes because of how many school days I skipped.

Most of my time was spent hanging out with my friend Julian and his cousin Martin, and freshman year, Martin started smoking weed. I went over to his house a lot and even slept over quite a bit. He would often offer me the joint or pipe he smoked, and I would decline. Brainwashed by all the D.A.R.E. propaganda, I still thought weed was the devil's lettuce. I'd never been around anyone who smoked it, so seeing how it affected Martin was new to me. It didn't do anything to him; just made him calmer and more relaxed, nothing crazy, from what I can remember.

But then, one night, everything changed for me regarding my perception of the substance. I was spending the night at his place, and in the back of his house was a shed. He smoked a joint outside the shed, and after, we went in it to hang out. There was a school desk in the shed, so he sat down at the desk, and we started talking. At some point, while we were talking, he rested his head on his bicep. I remember talking and talking and him not responding, but his eyes were still open, and he was looking in my direction. Perplexed, I said his name in my normal tone,

"Martin," No response, so I spoke a bit louder, "Martin."

"MARTIN!" And suddenly, his head shot up, and he came back to reality.

"What happened man?" I exclaimed. "I was there talking to you, and you were just in a daze."

"Yeah man, sorry about that," he responded. "I was in my own little world."

That moment I thought, *wait a minute, wait a minute. Did he just say he was in his "own little world?"*

I took a few minutes to process what he said, and after some discussion, I decided to climb aboard the weed spaceship and experience my little world. It was exactly what I figured it would be: the temporary escape from so much pain in the real world. I loved it. There was something about getting high that resonated with me. Alcohol was something I had experienced before, though I had never gotten drunk up to this point. On the other hand, Maryjane had me from the moment we kissed.

From that point on, Martin and I were basically "best buds," and we would skip school to get high. Our school had off-campus lunch, so we would just leave and not go back for the rest of the day. That year I had about fifteen absences in the morning classes and thirty or forty in the afternoon. There were never any repercussions for skipping, so I didn't intend on stopping. It wasn't just weekdays, of course—our lives revolved around getting money for weed, trying to get weed, and then trying to enjoy it in peace.

At the end of the school year, I thought I would have to do summer school again or hear complaints from someone, but I don't recall that ever happening. Somehow, I managed to get through freshman year without catching hell for all my days missed, and that summer into my sophomore year, I went to visit with my aunt in Tennessee. I had seen her in the past, but this time was the best since it seemed the neighbor's kids had made the same transition as I did: from video games to smoking weed and drinking just about every day. I HAD MINIMAL SUPERVISION when I was in Tennessee because my aunt was a nurse at a children's hospital and worked most nights.

Out there, I thought how insanely these privileged kids partied; I'd never experienced anything like it. They were not only able to score weed, but they were also able to get alcohol from a convenience store consistently. Going in once to help get the alcohol, I'll never forget being blown away at the number of security cameras in that store. At the time, Memphis was the country's crime capital, so that was likely the reason, but holy shit, I was paranoid; walking into that store set off so many sirens in my head. But I played it cool, got the alcohol, paid for it, and got out of there.

I know we partied a lot that summer, so I still have only a few real memories of that time. The first thing I remember because I was sober was when we were trying to get some weed. We had just got back to the suburbs, but for some reason, one of the crazy guys in our car stuck himself out and started to holler at a car full of young black men that were driving next to us. We were on suburban roads, probably going forty, and they were screaming back and forth about weed. Suddenly, one guy in the other car behind the driver stuck his torso out the vehicle and said,

"I can score you pounds if you want it!" He gestured a significant amount by spreading his arms wide.

The vehicles swerved as they came closer together, and the force sent him back into the car, and we went about our ways from there. A friend of theirs had started the whole ordeal with the guys in the other car, and I remember my buddy Aaron being mad at him. He scolded him about doing something stupid and almost causing us to wreck.

I had many escapades like this around this time in my life, but one thing happened that summer that probably had some of the most profound implications on my life. One night, we went to a house party at one of their friends' houses. I started drinking, and it didn't take long for me to get drunk, as I wasn't yet used to drinking. A few drinks in, I got hot and took off my shirt, then I remember some bigger guy talking crap about me drinking "bitch drinks." I started mouthing off in my dumb drunken state too and walked towards him. That's when Aaron came in and stopped me.

"That guy's gonna kill you," Aaron said. "He's way bigger than you."

I tried to ignore my friend, but before the big guy could get his hands on me, Aaron took me outside, and I drank and smoked some weed out there. That's when things went south. The weed and alcohol mix made the entire world start to spin, and from that point on there are only flashes of memories. I remember leaving the party and going to someone else's house. That's when it was the most intense: the world was spinning out of control (though somehow I managed to not puke). From there, we went to a late-night donut shop, all of us loaded in one car, and I was sitting behind the driver's seat.

Once we got to the donut shop, all I wanted was a cup of water. My "friends" asked me to shell out some cash for the donuts. I think they wanted it as payment for babysitting me throughout the night. I handed the money over on the condition that they bring out a cup of water to help me feel better. They agreed, took my money, and went inside.

I waited, sitting in the car dying of cottonmouth from the weed and dehydration from the alcohol I had consumed. Meanwhile, they were in there having a blast. I was hanging out the window. That was when a teenage girl and her mom came out of the shop and got into their car parked a few spots next to us.

I don't remember what I was saying, probably complaining about my lack of water. They just sat a few parking spacing over talking with me from their car, not knowing what to do with a drunk kid sitting alone. At some point, after they left me, I decided, "Fuck these assholes."

That's when I started honking the horn hoping they would come out to quiet me down. I don't know how long I did it for, and they still wouldn't come out. So I honked some more. I kept laying on the horn and honked and honked and honked.

I eventually mustered the energy to drag myself inside to see why they hadn't brought out my water. I walked in, furious. "Where's my water?!" I cried.

"Dude, chill," Aaron quickly replied. "They're already threatening to call the cops cause of your honking."

Angrily, I told him that all of this could have been avoided had they just brought out my water like I had asked them.

The next day was horrific. I was extremely hungover and felt like death. It was then that I decided that I would never simultaneously smoke weed and drink alcohol ever again. In my mind, I thought I had to choose one or the other, and I probably thought about it for a millisecond. The choice was always a no-brainer: I would continue to smoke weed and not drink anymore. Weed was easier to get back home, cheaper, would last longer, and was more enjoyable. I could go on and on.

CHAPTER SEVEN

FORGET BRAWN, IT'S ABOUT BRAIN!

Sophomore year was a lot of the same attitude about life: I crossed most bridges only as I came to them. I decided I would play football, though I still planned on smoking weed and skipping school as usual.

With football, it wasn't even something I was particularly interested in playing with football, and I just went through the motions. As for playing time, I didn't get much, and it was a shame. Not that I deserved to be in there since I didn't give my all to anything. There was never anything that piqued my interest. In the end, though, my whole attitude was a shame because I could have been good. My eye-hand coordination is second to none, all thanks to the thousands of hours playing video games. I was great at catching and could've been a huge contributor to the team.

We were running a flood pass one time on offense, and I was the backside option. We ran the play, and my defender immediately left me, and I was running my route all alone. I thought the ball would be thrown my way as there wasn't a defender anywhere within a ten-yard radius from me. Unfortunately, the ball went to a primary route runner and I was irate.

After the play, I was pulled from and wanted the coaches to know what happened. I was trying to help the team win, and I figured the coach would appreciate knowing that I was open on that play. I went up to him.

"Hey coach, I was open on that play," I said, but there was no response. "Hey coach, I was open on that play," I said again. Again, no response or acknowledgment. Coach didn't even look at me, wave me off, or anything, so I thought he didn't hear me. I spoke up even louder.

"Hey *coach*! I was *open* on that *play*!"

With that, he snapped back at me and made me shut up. Though I didn't remember what he said, I thought, "What the fuck? I'm just trying to help the team!" The attitude from my coach made me a huge spoilsport, and at that point, I was pretty much done.

My time to shine that I had screwed up came on defense when I played cornerback. I rarely went in to play corner; the majority of my playtime, little as it was, came on offense. I answered the call and went in, and this time the guy I was guarding had the ball thrown to him. I think he ran just straight, and I was running with him, my head turned towards the quarterback. I saw the ball get thrown in my direction.

This is it; there is no way I'm not catching this ball, I thought as I saw it soar towards me. The ball sailed through the air for only a few seconds when I decided to stop running with my guy. I jumped for the ball. I timed the jump perfectly but came up just short. The ball had barely grazed my hand.

I was able to jump pretty high in high school. I was almost able to dunk a basketball and could sometimes dunk a volleyball, which isn't bad for an unathletic young man of average height. My jumping and ALMOST intercepting the ball had caused enough distraction for the receiver not to catch the ball.

As my feet hit the ground, I turned and saw the receiver with the ball laying next to him. I was relieved since we weren't too far from the end zone, and it would've likely ended up in a touchdown if he caught it. After that play, I was pulled out of the game, and a different coach told me I shouldn't have stopped running with the receiver, and I would have intercepted the ball. That was one of the only moments I was excited during the season. That was the only time the ball ever ended up near me in all my playing. I took a mental note not to jump early and just run the ball down but never got my chance again.

During the season I kept that bench nice and warm and the only joy I got from being on the team was during the free meals when we played out of town. Even then, though, I made way too many mistakes. One missed opportunity led to the next, and I ended up acting even more like a zombie, and I just went through the motions. During the season, I kept that bench nice and warm, and the only joy I got from being on the team was during the free meals when we played out of town.

Early in the school year, my current girlfriend Monica's mom had a little talk with me about smoking weed and skipping school during that football season. She wasn't going to allow her daughter to date a pothead. I explained that my friend Martin recently moved away, and he was the only one I knew who smoked weed, so it was easy not to be around weed anymore.

At some point, then, Monica's family became my family. Her dad worked in the oilfield, so he was able to provide a nice living for his family of five by doing that. I was with them all the time, and it was a lot like being temporarily adopted. From after school until dark and on the weekends, I was with them 90 percent of the time. I ended up close with the members of their family, and was like a big brother to Monica's younger brothers. I believe this was the first time I felt a genuine *family* feeling, and those relationships motivated me to try and finally do something with my life.

Monica's mom encouraged me to seek out skills and classes that would give me potential in a future job, so I enrolled in agriculture and FFA classes in that same year. I wanted to take as many of the FFA classes as I could in the lead-up to the shop classes, where I could do metal and woodworking my junior and senior years.

Mr. Johnson, or "Sly," as we called him, was the agriculture teacher, and he was superb at his job. He made learning exciting and fun, and the best part about the class was Plant ID Fridays. Just about every Friday, he would go and grab a bus from the bus barn and pull it up in front of the agriculture building. We would all pile in the back, and Mr. Johnson would drive. Our first stop was always a convenience store for some quick snackage; then we were off to random places around town to identify plants.

Growing up on the ranch, I had always been interested in the different plants and their strange names. My dad was able to tell me the names of the local wildlife like the rabbits, quail, peccary, and deer, but no one I knew could tell me the plant names, let alone show me cool things about them. So every Friday, I stuck close to Sly. I tried to learn everything about the plants that I could. I asked questions and listened closely to everything he had to say about the plants. It wasn't long before he noticed my interest and told me about the Plant Identification competition the FFA held.

Mr. Johnson said this contest was one of the hardest competitions as you had to identify hundreds of Texas range plants at any stage of growth and those plants' characteristics like annual/perennial, warm/cold, native/invasive. I didn't care about its difficulty; I was just interested in learning all I could about the plants, so I signed up. From that point on, passing my classes was mandatory because I couldn't compete if I failed. Though I had always loved learning, this was the first time in my life I cared about getting a passing grade.

I sure wasn't going to study and try to get straight As since the only studying I wanted to do was for the plant competition. I didn't do anything special to pass my classes besides trying a little harder to just get over that failing hump. But even after my new wave of dedication, learning about plants was the main thing I was concerned with. Everything else was secondary.

I quickly learned that Mr. Johnson wasn't lying about the difficulty of the competition, and there weren't too many other students that wanted to attempt it. The only two that took it as seriously as I did were two seniors. One was Sly's son, a tall, browned-haired smartass named Edward, and his best friend Calvin was slightly shorter with darker brown hair and even more of a smartass. It was likely that Mr. Johnson had been teaching Edward about plants for most of his life, Calvin being a close second. It was also clear that they both participated in FFA activities since elementary with the 4-H club. Either way, these two guys were exposed to this plant knowledge exponentially longer than me and I envied how much they knew.

Getting to know Edward and Calvin drove me even further to learn as much as I could to compete with them. They also needed me to compete since it took at least three people to make the team, with four being preferred, as they dropped the lowest score.

We went to a few smaller competitions throughout the school year and did well enough to punch our ticket to the state competition at Texas Tech University in Lubbock. With those competitions, our team was carried mainly by Edward and Calvin. Since I was so new, I was still doing a lot of work learning about the plants and getting used to what could be expected during competitions themselves. I remember my nervousness as the state competition was getting ready to start. I was standing in front of a plant positioned upside down, and once the competition started, we were supposed to flip it, but as I stood there, I felt like puking from the anxiety.

The first plant I had to identify was one called hooded windmill grass. Luckily, I knew everything about it. I scored my card and waited for the timekeeper to tell us to switch. After identifying that first plant, my confidence soared, and I blew through the competition. I felt good after it was over, like there was almost nothing I got wrong, which is virtually impossible in a Plant ID competition.

Afterward, we hit the road to get a head start on the seven-hour trip back home, and on the way home, Mr. Johnson learned that Calvin got first in individual points, Edward got fifth highest, and I got second highest. Together as a three-man team, we had won the state competition.

We rode high all the way home on our victory. Winning state in Plant ID with only one team of three was quite the feat; most schools had multiple teams of four at the competition. Calvin ended up getting a thousand-dollar scholarship to Texas Tech and a big country belt buckle for getting first place, and I got a five-hundred-dollar scholarship to the same school.

After our awards, Sly made fun of me for "being the first loser" by getting second place. He was just messing with me, trying to get me to be better for next year.

There weren't too many other FFA competitions I was interested in. Most of them were a little too based on a judge's opinion, and I didn't like that. I wanted any competition to have a black-and-white, right-and-wrong answer key. Plant ID was the only one I wanted to spend time learning.. I became the big plant man on campus after that because the seniors would soon be graduating.

CHAPTER EIGHT

A FAMILY LOST

My junior year was when Monica entered high school, and the tradition of seniors toying with freshman girls' hearts had begun. I warned her about the other guys that might want to mess with her, but she blew me off, which rapidly led to our break up.

The first few days after we broke up were hard because I was heartbroken that I had lost my "love," Shortly later, things got worse because I realized how much I missed her little brothers. I was their big brother, and we did a lot together. We played video games (of course), messed around outside and inside, listening to music, pranked each other, and did countless other things. I even went to their ball games; as I said, I was a member of the family right before we broke up. I ate dinner there just about every night and was with them right after school until the evening when my dad would come to pick me up.

When we broke up, it created a rift in my entire life. I no longer had a stable support system. However, I knew I had to focus on what I enjoyed, so I continued doing well in school, well enough to compete in Plant ID, anyway. I could also still play sports as well.

This year, my sport was basketball. I went into the sport confident since I thought I was a pretty decent streetball player. But I quickly learned that sanctioned games were foreign to me. I had zero concepts of the majority of the game's rules, had zero stamina, and had to translate the basketball plays as if they were an alien language. They made no sense: stack and pop? What the fuck is that?

Using our bodies to screen each other was something I wasn't comfortable with, given my history. I didn't want bodily fluids from my friends on me; even me touching them was a lot, let alone doing it to strangers. Just touching someone else in any capacity was a big deal to me, unless it was to fight. But those days were mostly behind me.

I don't even remember basketball practice too much because I was just so lost in it all. By this point, the coaches simply expected their players to be developed. In my mind, I was good enough to play varsity. I just didn't have the fundamentals or know the rules. It was so bad, I got called for three seconds in the paint and didn't understand what I had done wrong. I was just standing there like a big dumbass, thinking I could guard the hoop all day long as I could in streetball.

I started on varsity to start the year, primarily based on my streetball cred and my friend Jeffrey's recommendation to the coach. But soon, even Jeffrey quit the team, and I wanted to. I like to think the season would have gone a little better if Jeffrey had been there because I would've felt comfortable asking him questions. In retrospect, I should have quit the team when he did and saved myself the embarrassment of getting demoted from varsity for my lack of ability.

The worst part about the demotion was how it happened. The junior varsity coach had called me to his desk during class while everyone was quiet and working on an assignment. There he tried to ask me if I wanted to play on J.V. quietly so I could get more playing time. He didn't ask quietly enough, though, and I'm sure that some overheard our conversation in the classroom.

I wish I could have told Coach then and there that I just needed someone to take me under their wing and teach me the way Sly did with his disciples, but coaches who do those kinds of things are probably rarer than we like to think. Obviously, my lack of experience was the cause of my inability on the court, and was even pointed out one day when Coach was laying into all of us, pointing out all the reasons why we sucked.

Even on J.V., I couldn't hang. Even though I did get a bit more playing time, I had no stamina at all. We never played full-court street-ball, so the longest I had ever run nonstop was probably at most a half-mile, so when it came time to run up and down the court during games, I simply couldn't. During one game, Coach had to call a timeout to pull me out after I waved at him, gesturing to my chest. I felt like my heart was going to explode.

That was my horrific junior year sports experience, and that year would unknowingly mark my last athletic event in high school.

That year, for Plant ID, we had three to five people who stepped up, but we only took two with us again to the state competition: one of my classmates, and Martha, a sophomore. We took second place as a team, and I took second-highest individual points again, for two years in a row. Sly let me have it again, telling me I was the first loser. This was his attempt to motivate me to try and pull out the number one spot my senior year.

I had a massive crush on Martha, especially since we spent a lot of time together doing Plant ID. My crush on her was apparent, so apparent that one of her good friends asked me one day if I liked her or wanted to date her. A lot of guys wanted to date her, so to avoid conflict, I just said,

"No, I'm just happy to be her friend," which was true, but of course, I would've loved to have been her boyfriend. She was so much fun to be around. I never let on that I liked her because I knew I had zero chance. I figured, why even say anything and risk ruining our friendship?

One of the many times I was high at school she jumped at me out of nowhere and began moving her arms strangely in front of her face. I was tripping out because I was high. Like a ninja, she was gone as fast as she showed up. I found this interaction hilarious, mostly because she probably didn't know I was high.

Though I never dated Martha, I had a girlfriend from out of town for a bit. Jeffrey hooked me up with her. She was cute, but one day while at home, she called me, and with the wonders of Caller ID, I saw it was her and decided I wanted to keep watching TV. I thought it would ring a handful of times and end, but the phone rang, and rang, and rang. The phone rang for probably almost five minutes, and that's when I knew what I had to do. I realized I had to end our relationship as we had only been together maybe a month or two, and she was from a different town, so running into her was likely not going to happen. I broke it off and didn't have a girlfriend my junior year, but instead spent a lot of time with my friends to make up for the last two years spent with my ex.

This was also the year that I became noticeably bad at math, as it was the year that we took Algebra 2. I did well in Algebra 1 and Geometry, but shortly after beginning the year, our teacher resigned because of rumors she slept with our superintendent. As a result, they had one of the newer band directors teach us, and my friend Conner and I spent that entire year fucking off.

One time, while the teacher was writing on the board with his back turned, I flicked a piece of paper at him, trying to hit his back. I had missed, and he saw the paper hit the board and went into a frenzy.

He turned around and cried, "WHO THREW THAT!? WHO DID IT?"

Thankfully no one ratted me out, but I did feel a bit bad afterward about what I had done. We did very minimal work that year, and the teacher would pass us with Cs. We didn't learn anything, and that was the last math class I would take in grade school.

I never had much trouble with math, but plants always seemed to take my full attention. I continued passionately studying plants and preparing for next year's contest. Of course, my work for Plant ID wasn't unnoticed. After my "fame" from nearly winning the Plant ID competition twice in a row, there were other teachers who had tried to recruit me to join the academic competitions they sponsored. However, there was never a genuine approach from them. It always seemed like they just wanted to use me to their benefit to help them win the way I did with Sly in Plant ID, so I denied them all and just stuck to what I knew and loved the most.

After learning about them, identifying plants is all I wanted to do.

Even though I loved working with plants, the time approached when I would start studying for a trade. Junior year was the first year we were allowed in the shop to do wood and metalwork, and this year I had chosen to build a barbecue pit to enter in the local county fair. My dad got a tractor, and we brought a huge piece of pipe, over fourteen feet long and three feet in diameter, to the shop.

That year, Sly taught me how to be a great welder. According to him, the ultimate test was to layer weld beads on a four-by-four inch plate by welding over your own welds. Essentially, I thickened a ⅛" plate into a ¼" plate. He would then take that plate, chop it in half with a metal saw, and see how many imperfections were visible from being exposed by the cut.

The first time I did this, I passed but had imperfections. The second time, I took my time to ensure that I was grinding away all the fouling (residue) from welding with a wire brush head. I spent so much time meticulous about the angles, cuts, welds, and grinds. It probably took me an extra two or three hours per week doing it this way.

Of course, I got hate from others saying I wasn't a good welder, just good at grinding. But when we cut it, the inside had a mirror finish; it looked amazing! Sly was impressed, and that training he ran us through laid the foundation for me to make one of the best barbecue pits he had ever seen.

After months of work, the barbeque pit was complete. It took about six to eight of us to load it in the back of a truck to take it to the fairgrounds. I was sure that I would win grand champion that year, mainly because my project was done at school. Students had the option of doing their projects at home, of course, but Sly said he wouldn't vouch for their authenticity of being done solely by the student.

An ex's sister had also entered the fair, and she entered a crocheted blanket very similar to ones her mom had made in the years prior. I knew with 100 percent certainty that her mom made the blanket or at least did the majority, if not all, of the work. I ended up losing to that blanket and made only $300 on the sale because the reserve champion isn't allowed to sell for more than the grand.

I was so upset, mainly because I was sure that she didn't do the work herself. Had I won, that pit would've sold for thousands of dollars.

Reflecting on it, I wonder how a crocheted blanket that someone made at home would ever beat a huge, phenomenal pit made at school. My dad and his brother had a lot of enemies, so it's possible that someone could've pulled some strings to spite me. That may seem like conspiracy talk, but my life so far hadn't been working out too well, so suspicion had become a part of my nature.

However, the worst part about the pit situation was that I was fucked out of a lot of money, and it sucked because we needed that money as a family. We always needed it since Dad always gambled, drank, and snorted all our money away.

The lack of money in my family kept me in anxiety mode. I never even really went to the fairs because of it. I even got out of showing a sheep that I raised years prior since nothing would get me to stand in front of people, not even the chance of winning hundreds of dollars. Smiling and freely talking to people wasn't something I did, especially at this point in my life. I didn't feel like a human most days. Instead, I felt like that sheep who should just keep his mouth shut to keep from offending anyone.

At this point, I'd occasionally daydream about being happy, which was never something I used to do. One time, I daydreamed while we were riding to town in his work truck, both of us staring outside. I was happy and smiling for a minute, but I snapped out of it fairly quickly. The reality was that we *were* driving together, but we sure as hell weren't smiling. When we got to the baseball field area right outside of town, I came back to earth.

At this point, our reasonably new double-wide trailer was on the verge of repossession. Fortunately, Ruperto had recently bought another big house on a few hundred acres north of town, so we lived in his old home for a bit. I don't know where we would've ended up if we didn't have that place.

The house was in pretty bad shape. The living room floor was bowing from improper support, the roof was in desperate need of replacing, and one of the restrooms regularly gave us plumbing and draining issues. Those were just a few problems, and on top of no central air conditioning, things were usually pretty hot. Dad had a window unit AC in his room, but I usually didn't sleep with him as I had as a kid. Still, I was thankful for the roof and four walls protecting us.

While we were living there, I began learning how to drive. During that time, I somehow convinced my father to get me a red V6 sports car, even before I got my license. I didn't necessarily have that particular car in mind; I just wanted some kind of ride to get me around. All I knew was I wanted some type of sports car and that I was enraptured by the recent customized car culture that had swept the nation. For me, it wasn't just the movies—it was video games, magazines, and television, too. I never raced my car or did anything too crazy because I didn't want anyone to suspect I didn't have a driver's license. It wasn't worth the risk.

But of course, we got behind on payments shortly after getting the car, and my dad's excuse for not paying it was that it wasn't a truck, so he had no use for it. In reality, we simply had no money. His paychecks were likely going to other people. We had no mortgage or utility bill because we got water from a well and used burn pits for our trash. All we could pay for was food, light, gas, and that car, but only for a few months.

This was the year I started feeling the pressure to look out for myself and take steps to make money soon. I took the ACT and Armed Services Vocational Aptitude Battery test (or ASVAB). The ASVAB is the test required to join the military and is used to determine what types of jobs Uncle Sam'll offer you. Though my life wasn't great at this time, I still did my best to work toward my goals. I planned to go to community college in Corpus after graduation, even though I didn't need to take any test to get in. I took the tests and did reasonably well on both. I scored a 21 on the ACT with no preparation and no algebra ability and scored 72 out of 99 on the ASVAB.

In our school, there was a special section on a wall for naming students who scored a 23 or higher on the ACT. There were only a handful of students from all of the classes in the top ten who were good enough to make it on this wall. As much as I hated taking pictures, I wanted to be on that board to show everyone that I was one of the smartest in school, despite my reputation as a stoner student who missed almost as many classes as he attended.

So, I waited a year to try the test again. However, life wasn't about to get easier for me.

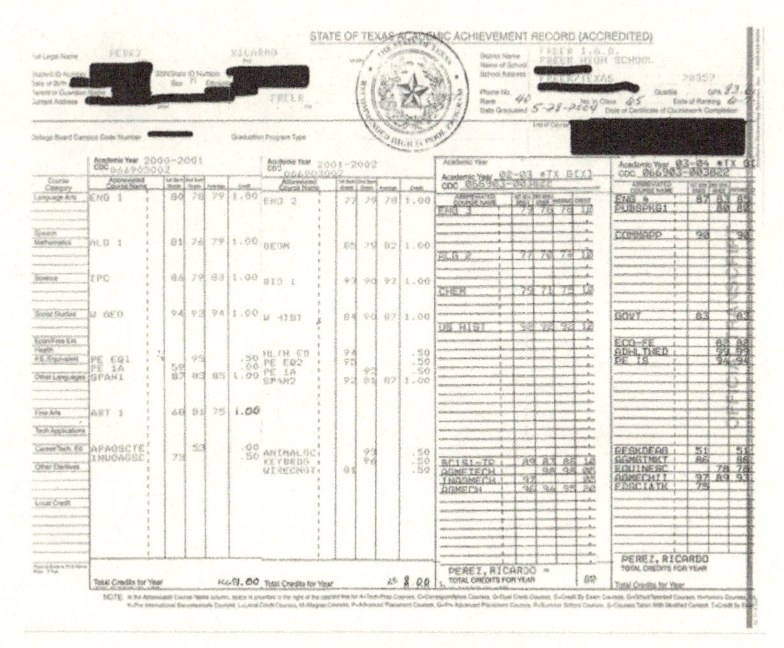

STATE OF TEXAS ACADEMIC ACHIEVEMENT RECORD (ACCREDITED)

Full Legal Name: PEREZ RICARDO
Current Address: FREER
District Name: FREER I.S.D.
Name of School: FREER HIGH SCHOOL
School Address: FREER/TEXAS 78357
Rank: 40 No. in Class: 65 GPA: 83.[illegible]
Date Graduated: 5-28-2004

Course Category	Academic Year 2000-2001 (CDC 066903002)	1st Sem	2nd Sem	Average	Credit	Academic Year 2001-2002 (CDC 066903002)	1st Sem	2nd Sem	Average	Credit
Language Arts	ENG 1	80	78	79	1.00	ENG 2	77	79	78	1.00
Mathematics	ALG 1	81	76	79	1.00	GEOM	85	79	82	1.00
Science	IPC	86	79	83	1.00	BIO I	93	90	92	1.00
Social Studies	W GEO	94	93	94	1.00	W HIST	84	90	87	1.00
Health / P.E.	PE EQ1		95		.50	HLTH ED	94			.50
	PE 1A	59			.00	PE EQ2	95			.50
						PE 1A		93		.50
Other Languages	SPAN1	87	83	85	1.00	SPAN2	92	81	87	1.00
Fine Arts	ART 1	68	81	75	1.00					
Career/Tech. Ed.	APAGSCTE		53		.00	ANIMALSC		93		.50
	INUDAGSE	73			.50	KEYBRDG		96		.50
						WIRECNGT	81			.50
Total Credits for Year					8.00					8.00

Academic Year 02-03 *TX B[illegible] (CDC 066903-003022)	1st Sem	2nd Sem	Average	Credit
ENG 3	77	76	78	1.0
ALG 2	77	78	78	1.0
CHEM	79	71	75	1.0
US HIST	92	92	92	1.0
SCISI-TR	89	83	86	1.0
AGMETECH		98	98	.5
INDGMECH	97			.5
AGMECH	96	94	95	2.0
PEREZ, RICARDO — TOTAL CREDITS FOR YEAR				8.0

Academic Year 03-04 *TX G[illegible] (CDC 066903-003022)	1st Sem	2nd Sem	Average
ENG 4	87	83	85
PUBSPKGI		80	80
COMMAPP	90		90
GOVT	83		83
ECO-FE		82	82
ADHLTHED		99	99
PE IS		94	94
RESKDEAG	51		51
AGMGTMKT	86		86
EQUINESC		78	78
AGMECHII	97	89	93
FDSCIATK	79		

PEREZ, RICARDO — TOTAL CREDITS FOR YEAR

Graduated 40th out of 65

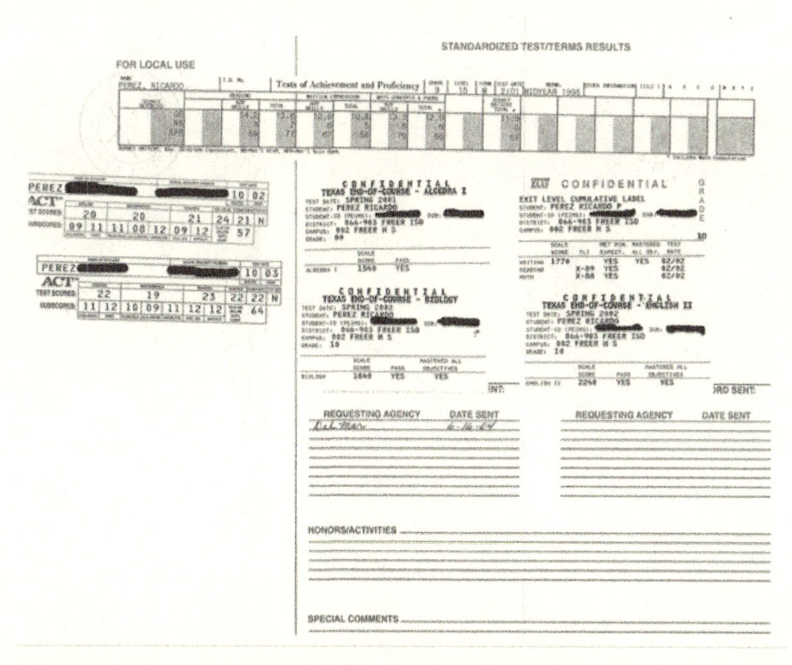

STANDARDIZED TEST/TERMS RESULTS

FOR LOCAL USE

PEREZ, RICARDO — Tests of Achievement and Proficiency

	English	Mathematics	Reading	Sci. Reas.	Composite	
ACT Test Scores (10/02)	20	20	21	24	21	N
Subscores	09 11	11 08 12	09 12		57	

	English	Mathematics	Reading	Science	Composite	
ACT Test Scores (10/03)	22	19	23	22	22	N
Subscores	11 12	10 09 11	12 12		64	

CONFIDENTIAL
TEXAS END-OF-COURSE - ALGEBRA I
TEST DATE: SPRING 2001
STUDENT: PEREZ RICARDO
DISTRICT: 066-903 FREER ISD
CAMPUS: 002 FREER H S
GRADE: 09

	SCALE SCORE	PASS
ALGEBRA I	1540	YES

CONFIDENTIAL
EXIT LEVEL CUMULATIVE LABEL
STUDENT: PEREZ RICARDO P
DISTRICT: 066-903 FREER ISD
CAMPUS: 002 FREER H S
GRADE 10

	SCALE SCORE	TLI	MET MIN. EXPECT.	MASTERED ALL OBJ.	TEST DATE
WRITING	1770		YES	YES	02/02
READING		X-89	YES		02/02
MATH		X-88	YES		02/02

CONFIDENTIAL
TEXAS END-OF-COURSE - BIOLOGY
TEST DATE: SPRING 2002
STUDENT: PEREZ RICARDO
DISTRICT: 066-903 FREER ISD
CAMPUS: 002 FREER H S
GRADE: 10

	SCALE SCORE	PASS	MASTERED ALL OBJECTIVES
BIOLOGY	1640	YES	YES

CONFIDENTIAL
TEXAS END-OF-COURSE - ENGLISH II
TEST DATE: SPRING 2002
STUDENT: PEREZ RICARDO
DISTRICT: 066-903 FREER ISD
CAMPUS: 002 FREER H S
GRADE: 10

	SCALE SCORE	PASS	MASTERED ALL OBJECTIVES
ENGLISH II	2240	YES	YES

REQUESTING AGENCY	DATE SENT
Del Mar	6-16-04

REQUESTING AGENCY	DATE SENT

HONORS/ACTIVITIES

SPECIAL COMMENTS

ACT composite scores on the left and state standardized testing on the right

CHAPTER NINE

MY FIRST LIFETIME ENDS

Senior year, things somehow started better than expected. I was happy to be a year away from graduation and motivated by the thought of finishing school and leaving the town for good. I had set up my schedule to do a few regular classes in the morning and then work in the shop all afternoon. I just wanted to work with Sly.

Debt collectors hadn't repossessed my car yet, so I still had a way to get around. This helped because I had to drive to the next town to see my girlfriend. We did a few things together over that summer, and once I started school, she thought it was better if we took a break from each other. It hurt me at the time; I was confused because things seemed pretty good on my end, but we hadn't been dating too long, so it wasn't anything close to the devastation I felt with Monica.

It wasn't long before I drew the attention of a freshman girl, Emily, though I was unsure about dating her at first. I thought she was beautiful, but I didn't want to be the guy I warned Monica about when she entered high school because of our age difference.

When I sat down and thought about it more deeply, I realized I wasn't sure if our maturity levels would match. Mentally, I felt I was far more mature than most peers. I also didn't want to go through another difficult relationship with a high likelihood of breaking up. However, my thinking got the best of me because she was just too beautiful for me to say no to. Because she was part of a divorced family, she didn't have any brothers, which allowed me to avoid a lot of drama. This was a significant factor in my decision to give it a try.

Things went well for the first few weeks, but quickly enough, she started to push me away. We pretty much broke up, though she let me visit her occasionally for a while.

Since Monica, I was always on the lookout for a girlfriend who could offer me a new second family—one that would take me in as hers did. So when I started dating Emily, I wanted something with her that I couldn't have. Breaking up with her, then, hurt like crazy because there was nothing I wanted more than to just spend time with her, while she often ghosted me and at the same time expected me to go wherever she went. I don't blame her, though; I was a mess, and she was too young to know how to handle the kind of person I was at the time.

She continued to act like she was interested in me but then backed away once I came running. The psychological toll she was putting my teenage brain through was monumental. So much so that when showing up at her house one evening, I decided to self-inflict to show her that physical pain was nothing to what she was doing to me mentally. I made this decision because I had enough of talking like nothing was wrong. This was when I thought I had enough and took a razor blade, held my forearm out, and began to slice in the same general area.

"YOU SEE HOW MUCH YOU HURT ME?!" I screamed. "I DON'T FEEL THIS BECAUSE I'M HURTING WORSE INSIDE!"

I hoped that would get to her. Once she started crying and begging me to stop, I did. But after we both calmed down, we talked about how much of a fool I was for doing what I did while she got a towel to help stop the bleeding. At some point, the bleeding had stopped, and I still didn't convince her to stay with me.

My interest in girls wasn't the only thing I lost interest in. Because I didn't have anyone pushing me to keep working, I was the only one convincing myself of what was important and what wasn't. I set myself up for failure and paid the consequences severely. I didn't even do Plant ID that year because Sly said he didn't like my attitude. My response, though, was to stay hardened. I didn't want to put the work in to study the plants. I excused myself by saying I had enough to deal with and felt like I was already as good as I was going to get.

I was also not allowed to work in the shop for six weeks because of my attitude, so all I had as an activity was a stack of busywork. Sly again tried to get through to me. He pulled me aside one day and stared down at me like an angry parent.

"You know," he said. "People don't care as much about your issues as you think they do." To me, reality spun around Emily. It was stupid to think so, since I couldn't even have her. All I did was shrug back—no desire to change, no desire to impress the man I used to look up to as a father figure.

I was allowed back into shop class at some point, and I ended up making another barbeque pit that year because my mom's ex-boyfriend wanted one custom-built. I sold it to him for a thousand dollars, which was the most money I ever made myself. The pit wasn't my best work since losing to a plagiarized crocheted blanket made me less motivated to put effort into it. If money was involved, I couldn't be too careful.

With $350 of that thousand, I bought a pound of weed. It came straight from Mexico, and lasted me pretty much my entire senior year. I sold very little of it—maybe 250-worth—to my friends during the year, but I smoked the rest myself. It was a real luxury to avoid the hassle of hunting down bud anymore, so I viewed this as an excellent investment for me at the time. It raised my spirits, and the night it came in was one of the most incredible nights of that year.

That night, we eagerly waited in town for it to arrive, and when it did, it was transported to my room, where we decompressed the brick. It fluffed up more than I was expecting, and there were some of the most beautiful buds I had ever seen. I rolled joints of straight hair, and that night, we smoked into the morning with no peak in our high. We smoked in the Executive, Ruperto's old broken-down RV with some couches and chairs we would blaze on.

A few weeks later, I made potent brownies with about an ounce of the bud and took the brownies to school to ingest them during health class. By the end of class, they kicked in. Big mistake: next on my schedule was PE.

We were playing basketball, and I was so high I played like Steph Curry, dropping three-pointers like nothing. I didn't miss a single one. There was just something about the high I had going that had me in sync with the ball and hoop. Then, in the heat of competition, I went up for a rebound and came down on Jeffrey's foot, severely spraining my ankle. I immediately hit the ground in pain, holding my ankle and yelling expletives.

"Stop cussing, Rico," Coach told me as he leaned over from his table to see what the problem was.

"My bad," I responded. "I'm in a lot of pain."

It hurt like hell. It was the worst ankle injury I had ever had, and I still had half a school day to deal with it. It completely ruined my day *and* my high, but if it weren't for the intense high I got from those weed brownies, there was no way I would've made it through that day. I ingested over 400mg of THC as a 150lb teenager, was in another dimension and had no motivation to do anything.

After PE, I remember going to home economics class when I was supposed to be in shop class. We were there to make some burger patties for them to sell at a fundraiser, and I didn't want to help because I was super high and in a lot of pain.

Sly noticed half of what was going on, so he told me, "You don't need to move your ankle to make patties, so just sit there and help." I listened to him, but I still did my work at a snail's pace.

After school, I limped with my friend Conner to his house, about a mile away. We moved like sloths due to my injury, and it was a miracle that we made it there before dark. My ankle had swollen to a softball size by this point, and I called my dad to take me home. The only time I had seen a doctor in the last seven years was when I had my migraines, even when I was on the brink of death from the flu, so I just had to go home and lick my wounds.

At this point, I started realizing a chain reaction: one thing after another slowly shattered my world. First, the pound of weed didn't even last me the entire year. It was gone in about six months, and I had gotten used to the kick it would give me. Problem was there wasn't any way for me to get more. LIFE GOT HARDER when I ran out, and I moped even more than usual. My girlfriend situation also caused me much stress, and I was always so angry that I lashed out regularly at others, including my closest friends.

It got to the point of me being so down in the dumps that my dad *gave* me a sack of weed. I was sitting on the couch, passively watching TV in my depressed state, when he came up to me. I'll never forget it.

"Here, son," he said almost sheepishly. "I think you dropped this." That was when he tossed a sack in my lap. I can't recall how I responded, but I was elated that I had some weed to smoke. It made me grateful that Dad sometimes tried to be a dad, however misguided his efforts might have been.

During this year, I spent a lot of time with my friends Malcolm and Conner at our friend Alonzo's house. We would usually get baked and play an evolved combat game all night. Those nights always managed to cheer me up. Alonzo had already graduated high school and was back from his first-year college experience, so he was usually available to blaze and hang out. We would usually get baked and play an evolved combat game all night. Those nights always managed to cheer me up. Before getting to know Alonzo, I had never played multiplayer before, and this was the first time I was asked to come up with a "gamertag" or moniker for my character. After sitting around for what seemed like thirty minutes, the perfect name dawned on me, and Bigballs36 was born. The number thirty-six came from my love for rap music.

During this time, I spent a lot of time at Ruperto's rodeo arena. Dad had a key to the side door for the arena next to the pool tables, and I would often ask for the keys so my friends and I could play without spending any money on the tables at other game rooms in town. Many good times happened at those pool tables, but there was one time we came close to getting caught by the police while smoking weed.

My friend Malcolm had an old truck with glass packs, so it was loud, and usually, he would drive up, do a donut or two, then park and come in. Someone had moved into small quarters in the back of the arena for whatever reason. When he heard the noise of the truck outside, he gave the police a call. We didn't think anything of Malcolm's antics at first, of course, as it was something he would typically do. I remember scolding him a few times, though, for bringing unwanted attention to us.

Within a few minutes of smoking a joint, the back arena door that we kept slightly cracked was thrust open by a police officer, who immediately asked questions. I calmly explained to him that it was my uncle's property, and I regularly got the key from my dad. The officer then told us about the call they received for the loud noise, and I apologized while assuring him that it wouldn't happen again.

A few weeks later, when I was driving home from school, my dad's old black truck that he cherished was parked away from the garage area. It looked a little strange when I saw it as I drove up, and as I got closer to it, it was evident that it had been purposely set ablaze. I went inside to see what happened, and when I walked in, my dad was sitting on the couch smoking a cigarette.

"Dad!" I hollered to get his attention. "Dad. What happened to the truck?"

"Someone set it on fire," he said casually. "Motherfuckers tried to blow it up. They had a rag in the gas tank, but thankfully I was on fumes."

He took a pull of the cigarette and then continued, "I was asleep on the couch taking a nap after work, and the dog started barking like crazy. I thought maybe she saw a hog or something across the fence, so just stayed laying down and didn't think much of it until I started to smell smoke."

I couldn't believe it. What in the hell did he do to someone who would be ballsy enough to come and blow up our truck in broad daylight?

Needless to say, I didn't feel very safe anymore. Not that I ever felt truly safe since I don't think I'd experienced that feeling in my life yet. But at this point, I lay awake most nights, worried. I often wondered if whoever blew up our truck would return at night and try to burn the entire house down while we slept.

I was so depressed with my situation in life that at some point one evening, I was just sitting on the couch I had in my room, alone in the house, as usual, pistol in hand. I was done. I had had enough. I knew that if I ate a bullet, the only person that might shed a tear for me would be my mother. I was close to doing it, but the thought of Mom's distress was what ultimately stopped me. But man, that cold steel felt so good against my temple. I remember wanting so badly to just end the torment of bullshit I dealt with daily.

I sat there for a while and thought long and hard about it. More of me wanted to end it than wished to stay. At that point, there seemed to be nothing to live for since I placed zero value on my life. I saw no future for myself and just wanted it all to end.

When I put the loaded pistol down from my head, it didn't make me feel better. If anything, it made me feel more pathetic. But that was a feeling I was familiar with, and somehow I went on without going back to sitting with a pistol to my head.

A few days later, I practically forced myself to attend high school graduation. I didn't want to attend, but I did. Looking back, the event was something I could've done without. We were also trying to take pictures with my sister, and they started bickering over petty shit. Dad and Mom still acted like teens with a vendetta when in each other's presence. At this time, he was still an addict, and she was still sour from it all.

It got to the point where my sister had to intervene. "Guys," she yelled out. "Cut it out! You're embarrassing him!"

At that moment, I realized that we, the kids, usually had to tell them how to act by themselves. When they were with each other, though, it was so much worse.

And that was the end of my school life. I was done after throwing my hat up with my classmates. It was time to move on with my life and go on a journey that only the universe could have scripted.

All the trials and tribulations I had been through were preparation for something even more dissident and destructive. Being born to fail wasn't an option in my world, and failure always seemed to be knocking at my door.

CHAPTER TEN

COULD'VE GONE SO WRONG

Shortly after graduating, I moved to my sister's apartment in Corpus Christi and enrolled in a community college. This was the first time in my life that I felt challenged academically. I realized pretty quickly that I had no study habits whatsoever and that I'd spent my entire life just getting by, doing the bare minimum. All I did was pay attention during class through high school, and that method helped me pass with ease. It was different in college, though: class was a fraction of the time, and they expected us to go and study AT LEAST three hours for every one hour of enrollment.

I didn't attempt to change my habits the first semester, which got me two Cs and a B. I also worked full-time as a bagger at a supermarket chain in the Texas area. It was a rough job for me at first because of my personality; I just didn't do the talking or smiling thing that management expected of me.

My interview there was probably the worst of it.

"You know," the HR lady said. "It's okay to smile during an interview."

I was nervous as it was with the interview. But sitting across from an extremely young and attractive lady with a low-cut blouse didn't help matters. Regardless, I got the job and started at $5.50 an hour, which was Texas's minimum wage at the time. I hardly had any bills to pay: just my cell phone, weed, and a different older model sports car. Mom and her new husband helped get me that one since I needed a ride to get to work.

My pay was barely enough to get by. I didn't last long with my sister; we constantly fought over the space and who was supposed to do which chores. I was soon desperate to leave and heard that my friend Julian lived in a two-bedroom, one-bathroom apartment in a suburb called Annaville. He lived there with his cousin and a couple of our mutual friends. It was a shady complex, but I begged them to let me live with them and stay on the couch. They let me. I threw them what money I could every month and did my best to keep myself out of their hair. During that time, I felt horrible imposing myself in a place that was already crammed, especially since none of them smoked weed like me, but I had nowhere else to go. Mom was still in Alice, Dad was in Freer, and my sister was batshit crazy.

Conner would come to visit us pretty often, and one night when a few of us were getting back to the apartment complex, a few strangers approached us outside. We had seen them before and always felt weird around them. This time, they approached us.

"Hey, you guys like pills?" they said. "We got some if you're looking." They were trying to sell us opiates, more than likely. As soon as they spoke, we were suspicious and told them to get the hell away from us.

After being inside for around an hour, many police rushed the complex and took the strangers into custody. We were happy that law enforcement had done their jobs and got them out of there, but it wouldn't be long before law enforcement also put us through the wringer.

A few weekends later, I had run out of weed. My regular connection had run out since he was just a fellow college student who would just buy a quarter pound at a time to have free smoke. So since I had run out, Conner and I thought it would be a good idea to go and look in our vehicles for roaches (the butts of previously smoked joints) or anything we could scrap together to smoke.

I managed to get a few scraps together and had it in the plastic to a cigarette pack in my right pocket. I continued digging around my vehicle while Conner was at his when suddenly, a guy came up behind me.

"Get out of your car and let me see your hands!" He cried. "I'm a police officer." I turned around, and a guy in civilian clothes was standing there. "What are you looking for?" he demanded.

"My keys," I said quietly. I knew it was unlocked, so I left them in the apartment.

"Oh yeah?" he said. I knew he didn't believe me. "Put your hands behind your back."

I chose not to comply; this guy just said he was an officer but hadn't shown any proof. I just stood there for a second, but all of a sudden, a police cruiser pulled up right behind my vehicle. When the uniformed police officer showed up, I realized what was happening.

I turned around, and the undercover cop reached into my left pocket where I was, unfortunately, carrying my rolling papers. "AH HAH!" he exclaimed. "I knew you weren't looking for your keys. Where is it?"

"Where's what man?" I said. "I don't have anything but those papers, and they're not illegal to have."

The undercover officer grabbed me by the arm and led me to the police cruiser. He told me to put my hands on the hood. I did, and he then took the uniformed police officer to the grassy area in front of where my car was parked. He pulled out his flashlight there, and they began searching the ground.

"I saw him throw something over here," I overheard him say. "Take a look."

This whole time, I couldn't believe it. I was scared, so I took a glance at the officers and waited for the perfect opportunity. When they weren't looking in my direction, I took my right hand off the car, reached into my pocket, and dumped the plastic full of roach weed that the undercover idiot had missed. They didn't see the move, and I thought I was saved.

All the while, my heart was pounding. I didn't think about it at the time but had they found *anything* in that grassy area; they would've pinned it on me. I guess it was better he imagined me tossing something over to the grass rather than imagining me pulling a weapon on him. There were way too many ways for this to have gone wrong.

After a failed search in the grass, the uniformed officer came over to question me. Again, I told him we didn't have anything and weren't doing anything wrong because we weren't. This went on for a while when the uniformed officer finally seemed to have enough with my bullshit.

"Come on man," he said. "Tell me the truth. The truth will set you free."

When he said that, I somehow thought it would be better just to listen. "Okay, fine," I said. "I smoke weed, but it's just me. You can drug test any of my roommates, and they'll pass."

He then asked if we had anything in the apartment, and I told them nothing besides some seeds, stems, and a pipe. Then he asked if they could go inside the apartment, and I agreed.

Once inside, they realized that the weed was the least of their problems. Our apartment was filthy, and after they saw that I had nothing worth spending time on, they pulled me aside.

"Look man," one of the officers said. "This place is a mess. And since you're not on the lease, I'm going to have to ask you to leave by tomorrow and not come back to this place."

That's exactly what I did. I helped the guys, and we cleaned all night into the morning. The officers were quite impressed with how much we got done over the sleepless night.

With everything packed in my car, I went to my sister's apartment to explain what had happened, and she let me stay with her in the meantime. We both knew it was temporary because, of course, we weren't going to get along for long.

I needed more money, so I was thankful that I asked for a raise at work a few days before all this going down. I told the manager that I needed more than minimum wage, as it wasn't enough to live on. I asked to be placed as a cashier instead of a bagger since they made about $7.25.

Like most days, I decided to smoke a joint as I drove to work. When I arrived, I found some keys on the ground. Being the good samaritan who I understood myself to be, I picked up the keys to take them inside to my manager.

Then I remembered that I should probably wash my hands before handing him the keys; otherwise, he would smell the resin on my fingers. And so I went to the restroom. Instead of washing my hands, though, I just hit the sanitizer, thinking it was going to be enough. Then, I went to turn in the keys. In my stoner state, I put the keys right in front of his face, my hands unnecessarily close to him. He grabbed them from me and said thanks, and I went about my duties.

Later as I was bagging groceries, I realized that I could still smell the resin on my hands. There was no doubt that my manager could smell the aroma, too. That was the moment I knew I'd fucked up.

During one of my breaks that day, he came into the break room.

"Hey Rico," he said. "You want that raise? There is a night stocker position available, and it pays $9.25 to start." Before I could utter anything, he continued, "But, you'll have to take a drug test."

"No problem!" I replied. "I would love the opportunity." I attempted to sound nonchalant.

After he walked out, I waited a bit and then got up in shock. I knew that they didn't want to make me a cashier because I was sure as hell not going to be cheery and smile at the customer as they wanted. I didn't care though. I was pretty much doubling my pay, and that meant I was going to be able to get a place of my own or with some roommates. First, though, I had to pass that drug test.

I was required to take the test within a few days, so I went to the only place I knew to get stuff to help me pass: The Magic Mushroom Shop. It was a small place filled with pipes, papers, and supplies for anyone looking to get something outside the norm for their smoke experience. I looked at their options that would help me pass my test, eventually settling on a product in a pill form.

On the day of the drug test, I followed all the directions to a T. I was pretty nervous when filling out the paperwork and was thankful that the lady had offered me a sour apple lollypop. I filled out the information sheet, handed it in, and waited to be called on. It was a pretty packed place, and it took at least a half-hour for them to call me, but when they finally called me back, I was ready to go. As I began to relieve myself into the cup, I noticed a green hue. The pills somehow made that green lollipop come out in the sample. The color of my urine blew my mind.

I thought that for sure my sample was going to come back as positive; it was radioactive green, for crying out loud! I knew it was from the pills and candy. I could do nothing, so I handed the sample in and went home.

A few days later, I started working nights as a stocker. It took me a good month or so before I realized that my manager wouldn't fire me for failing the drug test. So I learned the job in no time and did it well, to everyone's surprise. I was proud to keep a relatively high-paying job for a high school graduate not working in the oilfield. With this raise, I quickly got into another apartment with my two friends, Alonzo and Malcolm. Overall, living together went well, considering we weren't living in an apartment with a room for each of us.

When the following semester started, I worked at the store all night from 10pm to 8am, sometimes for longer. I would then rush to school and spend time there until around noon. Then I'd immediately go home and sleep and do it all over again. It was just a matter of time before that shit got old, and in that time, we were also transitioning to a new store. When we moved to the new store, I was laterally "promoted" to the overnight dairy position. The job had me hauling and dumping out all the expiring milk and other dairy products. I wasn't happy with life, except with a new development: my new girlfriend, Eve.

I met Eve in the previous semester at an incredibly long and tedious student loan seminar I was required to attend. Eve was everything I wanted in a girlfriend at the time: wild and video game obsessed. We flirted with each other during the seminar, and she left before it even ended. When Eve got up abruptly, I was afraid that I'd never see her again. But not long after she left, she came back with her number on a piece of paper and told me to give her a call. I gave it a couple of days and tried to call, but she never really answered. I'd later see her again in one of the buildings on campus, and from then on, she gave things a shot.

Initially, she told me, "Don't expect too much. I usually get bored with guys pretty quick."

So, I didn't. We got along well and spent a lot of time together. I was done going to school and spent that time with her instead. I was also still occasionally talking to Emily, who still drove me crazy. I still hung on to her because I wanted her to come to live with me in Corpus, even though I knew that was unlikely to happen. After all, she was still in high school.

Though I finally felt I had some control over my life, there was no getting around the depression. I was depressed about my family, my breakup, and the stress of work and school. It eventually took a toll on me, which led me to get all Fs in my courses one semester. But I didn't feel like giving up yet. I took the Fs and decided to join the Army.

The three of our apartment occupants all enlisted; we had considered before. I went to the recruiting station in Kingsville: a single cold and depressing gray room. That was where I told them I wanted to join as an infantryman. I had two personal reasons for this. First, I wanted to fight because, in my depression, I didn't care if I lived or died. Second, I wanted to try and become a special forces soldier.

I signed up at the military processing center into the delayed entry program for four-and-a-half years. This got me a ten-thousand-dollar enlistment bonus and a thirty-thousand-dollar "Army College Fund" benefit. They told me I could take less enlistment bonus money and max out my ACF. Since it meant there would be more money in the long run, I agreed to the terms. Even though I wasn't in college, the recruiter had doctored up some documents from the local university to qualify me for the delayed entry program.

Even though basic training was a few months away, Alonzo, Malcolm, and I broke our lease at the apartment and went our separate ways. I ended up in Freer, back in my room where I had been depressed many times before. I wasn't sure if they were going to drug test us, so I stopped smoking and instead went for short runs, did pushups and sit-ups, and learned how to do the exercises in cadence with a DVD the recruiting station had provided. I would also spend time studying because if I knew things like rank and army values, I could go into basic with the rank of private first class with a bit of paperwork from my recruiter.

The months went fast, and in no time, I was on a plane headed for Fort Benning in Georgia, the home of the Infantry. Before we departed from San Antonio, they had us rush through more paperwork. Little did I know that this day and contract would end up adding fueling the fire after being discharged so many years later.

CHAPTER ELEVEN

THE GUT CHECK

Once I got to where I needed to be on Ft. Benning, they had us line up in rows and make a formation, and then they left us and went inside for some reason. We stood out there for at least thirty minutes (though it felt much longer than that) before they came back out for us and said they had forgotten about us.

The in-processing was supposed to take about five days. This was where we received all our equipment and inspections. They rushed us through there in two days, which was probably one of my worst experiences at basic training. We got little sleep, received junk like pizza and burgers to eat, and got a bevy of immunizations. It was entertaining to watch all the people with a fear of needles pass out or scurry in fear, though.

Luckily, at some point growing up, shots stopped bothering me. The awful thing about in-processing was how there was *never* any soap in any restroom. I got so desperate that I even risked going into female bathrooms, as there were no women around. I still couldn't find any. I was so mad because my face was an oily mess, and the anxiety of my situation compounded the problem. I tried to wash my face with just water, but what was previously my face had turned into a zit-infested oil slick, especially around my hairline, so the water did little to remedy the problem. There were substantial white pus-filled nasty things, and when they ran us through the buzzers for our haircut, they ran through the zitfield and busted them all up. Blood and pus ran down my forehead. The barber wiped away what he could and then just started throwing baby powder all over my forehead to try and clot everything up.

"Those sex bumps growing already, huh!" he laughed, referring to an apparent lack of intercourse.

After he was done, I went to the restroom to wash the powder off, but I was still stuck without soap. We went to our "company" to meet our drill sergeants at the beginning of the third day and got ready for training. They marched us over to the location in formation.

Shortly after arriving, I learned that since I signed up to attend Ranger School in my contract, I would be part of a platoon full of ranger candidates. Another platoon was designated as mortars, as we sign up as 11X and can end up as regular infantry 11B, or 11C, a mortarman. Mortars are an essential part of an infantry unit's operations as their indirect fire has frightening abilities. I was thankful that I had Ranger in my contract because ending up as a mortarman wasn't what I wanted.

After getting yelled at and the stereotypical basic training escapades, we each got one bunk and one wall locker. In my platoon were two others named Perez, and I was the middle. Then there was Big and Small. Big Perez was from the Kingsville area, so it was pretty cool that we were from the same region and got put in the same platoon. Small Perez was from California and liked to drink and party. Though he couldn't do it in basic, it wouldn't be long before he would show us the copious amounts of alcohol he was able to consume.

My first two-mile run in the Army was the first nonstop run I'd ever done in my life. I didn't think I was physically capable of doing it, and when I started the run, I was determined to keep running no matter what. I pushed through the horrible pain in my chest and side, and when I finally reached the finish line, I felt like I was on the brink of death. My heart pounded heavily in my chest.

I made it in nineteen minutes. It was far from the minimum required time for a two-mile run, but I had given it everything I had and not stopped. I never thought I'd be good at running, but to graduate infantryman training, you need to be able to run five miles in forty-five minutes, something I never thought I'd be able to do.

I didn't just experience pain from running, though. Throughout basic training, they ran and rucked us into the ground. We walked two miles on our first road march with about forty-five pounds on our backs. That walk was the most brutal and longest two miles I had experienced, way harder than my turtle-paced two-mile run. Everyone's feet hurt after that march, and we were dreading the four-mile march just a few days away. However, when we got done with the four-mile march, the consensus was that it wasn't nearly as tough as that original two-mile march was.

I kept my head down during all of basic training. Even the drill sergeant admitted to me towards the end that I had lucked out that they didn't pick on me more. When things got physically demanding, I would sing a rap song about collapsing in my head to get me through. The song starts with the rapper motivating the listener never to give up.

Shoving all my food down my throat in less than five minutes took some getting used to, and I never threw up. There were some close calls, though; I almost lost it due to others' puking.

My only physical confrontation in basic training was with a guy twice my size. He was hanging out on Big Perez's bed when he leaned over and started to eye gouge me with his thumbs for no apparent reason. As the pain quickly intensified, I reached for the dip in the center of his collarbone and shoved my thumb down in there as hard as I could. He released almost as quickly as he got his hands on my eyes, and it seemed like he didn't want to escalate it. Good thing, too, because I'm pretty sure he also had a wrestling background, and I would've had no chance.

The most fun I had in basic training was platoon pillow fights. Once we reached a specific week, there would only be one drill sergeant to watch all four platoons throughout the whole night, so once we knew the coast was clear, a different platoon on the same floor as us would rush over, and we would throw down in an epic sixty-man war.

To end basic training, we performed a required ten-day field training exercise. The ten days ended with a roughly ten-mile march to the ceremonial infantryman birthing ground. As I walked up to the holy land, other companies of trainees who were still weeks behind us in progress were there. They lined up their flashlights with blue and red lenses to cheer us in. We lazily walked through the crowd to get our cross rifles and officially became infantrymen.

"Man, these guys smell like SHIT!" I heard someone say as we marched in.

"Yeah," his buddy replied. "But it's the smell of victory."

I had a good laugh at that and then began to smell myself. I thought I didn't smell that bad, even after ten days with no shower and sleeping in a pit of dirt. However, after I'd had a shower, I smelled my old clothes, and they smelled distinctly like wet dog.

I was fortunate to go straight from basic training into Airborne School with no holdover wait for the class after me. After fourteen weeks of strict diet and training, it was awesome to have weekends off in Airborne School.

We usually partied hard on the weekends and only had to get a room on one weekend because one of my buddies from basic knew someone who lived in the barracks on base. We went over there pretty often, including during the week because he lived only a few hundred yards away from our place.

After two weeks of getting jerked around in a harness and a third week of jumping out of planes, I graduated Airborne School. I learned then that I was getting ready to go to RIP, the Ranger Indoctrination Program, because I had signed up for the Ranger position in my contract. This program was supposed to be another three or four weeks of hell on earth; the intention was to weed out everyone ready to give up or change their minds. However, they didn't need to do anything to get rid of me. I had realized that I didn't want to be in the Ranger battalion when I signed up. I just wanted to attend the Army's Ranger school. Here's how I saw it: if I was going to go through weeks of hell, then I was going to do it in North Carolina at Special Forces Assessment and Selection. Another reason I didn't want to get stuck in a Ranger Battalion because two out of the three battalions are in Georgia, and I wanted to be stationed somewhere completely different from where I was raised. In my mind, I wanted to go somewhere fresh and new like Italy or Germany, or perhaps somewhere further north like Alaska or

Washington. I didn't care where it was, as long as it was drastically different from anything I had ever experienced. After all, that's kind of why I signed up in the first place.

After voluntarily withdrawing from RIP, I waited around for about a week before getting orders to head to my new duty station as a qualified Airborne Infantryman. As they passed the list around, I got anxious to see where the Army would send me. The clipboard finally got into my hands, and I searched for my name. I was destined for Fort Lewis, Washington. I was ecstatic that I had got stationed at one of the places I had hoped.

A day or two after I saw where I was headed, my orders sending me to my duty station came in. I was off to Washington state and couldn't wait to get there! Little did I know that the unit I was going to, 5th Brigade 2nd Infantry Division, was an unstable neutron star on the verge of collapsing into a black hole, even though it had just been created.

CHAPTER TWELVE

BLOOD MAKES GRASS GROW GREEN

I got to Fort Lewis in October of 2006. When I arrived, they put me in some old building on the main post where we bunked with quite a few others. I noticed, then, that a few guys stayed with me since basic and airborne training, and now they ended up with me at Lewis, too. Even though it was only six months of TRADOC (also known as training hell), we endured. I felt like I had a new family. I would soon learn later that the brotherhood we built would last the next three years into our deployment, and we would depend on each other in so many ways, especially overseas.

In our living quarters, we had bunk beds. It was our staging point as soldiers because the 5th Brigade was so new that it didn't have buildings yet. They were some of the oldest buildings on post, other than the old World War II barracks housing units on the North Fort. I was assigned a room, and since I was one of the first privates, I was supposed to have a bunkmate. I didn't get one the entire time I was there.

When I went into my room for the first time, I found the whole floor covered with dried brown goo. I wasn't sure what it was at first, but as I mopped it up, it became apparent that it was days-old vomit. It was clear that men were already getting blackout drunk right from the start. I couldn't believe they would give me a room that had biohazards all over the floor, and it was just an omen of what was to come.

We didn't do much those days—primarily odd jobs until lunchtime. Then, we were usually free to go for the day. It was easy; we didn't do much but play games and head to the recreation center to play pool or other sports. We usually did physical training in the morning, though my lungs couldn't handle the cold air. I had never run in the cold, so even though I had become a pretty good runner, it took a while to get used to the particular sting that burns the lungs when you run in frigid weather.

Our first upper enlisted NCO (non-commissioned officer aka sergeant), Master Sergeant Miller, quickly made it known that he gave no fucks. He did whatever the fuck he wanted, and when it came time for the end-of-day formation, he would stand there in front of all of us and yell at us.

"If you think I'm mean, hurtful, mistreating you, I don't care," he would say. "Report me! Look, my name is right here." He then would pull on his uniform and point to the name on his chest. "M-I-L-L-E-R, that's how you spell it. Make sure you get it right so they can find me." He was throwing the red flag up to us the whole time he spoke; this let us know right away that there was nothing we could do to upper enlisted.

The power dynamic became more and more apparent as the unit got shittier and shitter sergeants. During this time, I had requested to take leave early to go to Texas and marry Eve. After conversing for hours on the phone and dealing with a long-distance relationship, we ended up agreeing that it was just something we should do to continue our relationship. We decided to behave as if we were just boyfriend and girlfriend, with no strings attached if it didn't work out. The plan was for me to fly down, get married, drive us both up, and get into an apartment pretty quickly. This all would happen within just a few weeks.

The road trip up was crazy; it was my first time driving through significant snowfall. I hit a patch of black ice somewhere in Idaho while driving a box truck hauling a car while Eve followed in my pickup truck. I flew off the road, and the only thing that saved that box truck from being totaled on its side was that trailer. The weight on the back kept the ass end from getting too crazy, and all that ended up happening was the vehicle flying off the road, hitting the dirt, and tipping over, with the weight of the trailer keeping it down. The bumper of the box truck ended up sitting on the hand crank for the trailer, in a somewhat jackknifed position. Thankfully, a big rig driver with no load stopped by, hooked up, and went bonkers to get me unstuck. He tried being as gentle as he could at first, but he came out of his truck after a few attempts.

"Well" he said. "I'm gonna have to do it harder for it to get unstuck. Do you care?"

"Absolutely not!" I replied. "I appreciate you even doing this; I'd understand if you just got in your truck and left right now."

"No way, I wasn't even supposed to be here tonight. I was supposed to have a load, but for some reason, it was canceled, so I had to come back this way and saw you fly off the road. God put me out here to help you, and that's what I'm going to do."

His kind gesture blew me away, and I was extremely thankful. I explained that I had never driven in such conditions and was a new implant in the Pacific Northwest. He went back to his truck to keep pulling, and this time he was violent with it. It was jerking his truck good, and finally, with some momentum built from incrementally rocking it by gassing it hard, the bumper snapped the trailer lift off, and the vehicle wasn't high centered anymore. From there, we were able to hit the road again!

Once we were back in Washington, we got a one-bedroom, one-bathroom apartment about fifteen minutes from base, and in no time, our workplace shifted from Main Post to North Fort Lewis. We had brand-new buildings to work in and barracks for the men to live in, with more of both still being built. The Army assigned us to the 1st Battalion 17th Infantry Regiment, a battalion that has been around since the Civil War. There is a medal of honor room in the main battalion building to honor the men who earned the prestigious award while in the unit.

The sister battalions were the 4th Battalion 23rd Infantry Regiment, 2nd Battalion 1st Infantry Regiment, 8-1 Cavalry, and another support battalion. We were constantly getting new men, weapons, and vehicles. Suddenly, what had started as a small group of soldiers turned into a full-blown unit with the capability of training for war.

As we got weapons and assignments, I hoped to be a squad-designated marksman because I knew I'd be good with a weapon system like that. I wanted to contribute the best I knew how. This dream quickly shattered when they made me a 240B machine gunner. I was furious. I at least wanted a rifle, something I've known how to employ since I was hunting as a child in Texas.

I WASN'T AS UPSET once I could get that thirty-pound (empty) beast to a range. Suddenly carrying a fully automatic death machine wasn't such a burden, and I proudly held the position as a machine gunner in the weapons squad of 3rd platoon. Every platoon designates their name, and we called ourselves the Dirty Pirates.

The weapon squad was the best because we had the best squad leader: former Marine Staff Sergeant (or SSG) Eddie. Thankfully I had him as a mentor. There are few men like him. Eddie was someone who cared about his soldiers and had the competence to lead them properly. He was a new Sly in many ways, but what he was in charge of teaching us and guiding us through was much more consequential.

The entire weapon squad grew close as we strived to be the best. The other machine gunner in the squad, Bidzii, was from the Navajo Nation in New Mexico. He was always smiling with a golden heart. Being fairly big and strong, having him around to push me helped me get stronger physically and mentally. It's also where I met one of my closest brothers, Duncan, a Washington local who grew up just a few minutes down the road from the apartment I was living in at the time. I couldn't believe that he grew up so close to where we were stationed and that the military kept him only twenty minutes from home. If I were in that situation, it would have driven me insane.

There were quite a few others who grew up in the area and ended up stationed there. But Duncan was bright, and that's why I think our bond was closer than most. He was one of the few that I could talk to about non-military topics. It was all about the military at any other time: we trained hard during the week, and I was usually asleep by ten o'clock because they expected us to be at work by 5:30 the next day.

One morning at our company area, they asked, "Who wants to try out for the sniper team?" A lot of hands flew up, including mine.

"Am I even allowed to do it?" I asked myself. I was hesitant because they spent so long training me as a machine gunner; why would they waste even more time letting me try out for the sniper team? I hoped they wouldn't deny me, now that they knew my interest.

They went on to say they were looking to fill three sniper spots. They summarized what the tryout would encompass, and a few hands dropped after hearing it. They wrote my name down and the names of about sixteen others, most notably Clint and Stephen, two of my friends from 3rd platoon who had combat experience and Ranger tabs. The other 90 percent of us who raised our hands were lower enlisted fresh recruits.

Clint had previously served overseas with the 2nd Ranger Battalion at Fort Lewis. Stephen had previously deployed with another regular infantry unit and then ended up at our unit after getting in trouble during the Special Forces Qualification Course.

We all knew they were likely to get two of the three spots, so the tryout was essentially for one position.

Shortly after offering us the tryout, they dismissed us, and I got excited. I would get an opportunity to prove myself for one of those sniper spots, and I would do everything I could to set myself up for success. I decided to start mentally and physically preparing myself for the tryout, even though the memorandum with expectations had yet to be released.

In the same 2007 summer, all training ceased. They tasked us out to help train the cadets in an annual event called Warrior Forge. Cadets from around the country go to Fort Lewis in the summer to "forge" into officers for the Army. Our days usually consisted of showing up at the company area for accountability, eating breakfast, and then rushing out to the training areas. They eventually set up places for us to stay out there overnight to be ready for the early morning exercises.

Duncan, myself, and another friend of ours, Kelley, would rotate out of a lane that had us stranded with a "broken-down" hummer, with one of us going off to get help, is what we told the cadets, but played sniper instead of staying together. We spent maybe twenty minutes every hour doing something involving the cadets, so that provided an opportunity to do push-ups or sit-ups in the off-time. Then, as the cadets arrived, I was supposed to act distressed because my buddy was missing, and then we were to come under sniper fire, where one of us would pop off blank rounds in the distance. Then the cadets were to react to the fire and secure the site.

One day, Kelley came up to me as I was working out.

"What are you doing?" he asked, arms crossed.

"I'm getting ready for the sniper tryout."

"Oh, don't bother," he snorted. "They're not going to pick you."

"Oh, no?" I replied, angry that he would make such an assumption.

"Yeah man, they're not. They only want guys with combat experience."

Regardless of my buddy's lack of faith in me, I planned on getting the remaining spot. I blew Kelley off and kept pushing, and when

I finally got the memorandum, I hit the books too. I'll never forget the memorandum: "The Alpha Company Sniper Selection course is established to evaluate soldiers that volunteer for the three vital sniper positions within the company. Each soldier will participate in a grueling multi-day event that will test their knowledge, physical fitness, navigation skills, perseverance, and individual character."

Not only were they going to test us physically, but they also planned to interview us and test us over everything, from basic infantry knowledge to theory behind propaganda and other facets of war. I downloaded PDFs and put them on my PSP to read while doing nothing at work or out at Warrior Forge. Smartphones were years from being invented, so I had to improvise.

Some of the guys saw me a few times and thought I was playing games, but then I'd show them that I was reading in preparation for the tests. I did everything I could in preparation to try out for that one spot because it was an opportunity to fulfill a childhood dream, and nothing was going to keep me from earning it.

The tryout was to be sixty-five hours long and held over three days, and we were the only company in the battalion who had tryouts for their snipers. The other companies just placed guys in the position, likely only based on PT scores or because of their aptitudes. But for my tryout, I had to work harder.

CHAPTER THIRTEEN

HEART OF A MAMBA

Finally, the day for the tryout had arrived. The room was plain and nondescript—every guy there had his head down and seemed ready for the challenge. When I woke up that morning, I was confident in my preparation and planned on giving it all that I had.

First, we were given written tests consisting of two hundred questions. That was all morning. Then we had lunch, and afterward, it was time for the interview. The first sergeant and platoon sergeants were way more people than I expected to have in the interview. I remember waiting in line for my turn to enter the conference room in the company and trying to keep my heart rate level since I was nervous as hell.

I walked in, stood with my hands behind my back at parade rest, and looked at each of them while they prepared for note-taking.

"Begin," the first guy on the right said.

Without a moment of hesitation, I started to recite what I had practiced over and over at home. It was going flawlessly, then suddenly I had a lapse and couldn't remember what I was supposed to say next. While I was talking, they were mostly looking down while writing notes, which made it much less nerve-wracking, but they all looked up when I stopped.

I quickly ran what I had previously said through my mind and remembered what I wanted to say next. I continued, and they all looked down almost in sync and began to write as I began to talk again.

From there on, it was flawless. I walked out of that room, holding my head high. After interviews there, we were off until 8:30pm, where we would participate in an all-night land navigation course into the morning; after that, we would run through a modified physical fitness test.

From then on, it was all hell. Everything in those next forty-eight hours was some of the hardest things I ever did in the military.

During the entire ordeal, I slept a total of four hours at most. I walked and ran forty to sixty miles, both with and without gear. As for food, I ate maybe three meals and threw up each of them, mostly from physical exertion. But still, I gave every event my all.

On the last day, we were all pretty physically beat down and soaked in our sweat from head to toe. We had just finished the twenty-four-hour mock mission. I think I had sprained my right groin by this point, as there was intense pain in that area as I tried to run.

For the final event, we were to run to the entrance of Solo Point, a spot on the Puget Sound near the military base. The easy part was running down to Solo Point; coming back up was a bitch. The run was two miles down the trail to the water and sea level, then another two miles back with an almost 400-foot stretch of elevation. It hurt so badly to run up that hill, but I did it for as long as I could. I'd occasionally stop to limp toward the goal line but not long enough for anyone to get close enough to try and pass me.

A few people who signed up didn't show up, a few others dropped out during the tryout, and nine of us made it to the end. I wasn't the first done in this event; my groin didn't allow me to get in any proper strides, but I did finish in the top 25 percent. And that was it; we'd finished the event.

After being released, the first thing I did was get to my truck and call Eve.

"Would you please make me some bacon and eggs?" I begged her. "I want some before I fall into some sort of coma."

All I got in response was a complaint. "Why can't you get something on your way back?" "I fell asleep just now," she said.

"I don't care if you just fell asleep," I said. I had just spent the last three days in hell, and all I wanted was that homemade breakfast. In any case, she fussed, I hung up. When I got home, there was a plate of food waiting for me.

I thanked her and apologized for being mean, but I was famished since I hadn't fully digested any food for the last three days. I'll never forget how good that first bite tasted when I put it in my mouth, followed by the horror of trying to swallow it. My throat had closed up some or something, and it hurt like HELL to swallow. I could feel the food working its way down my esophagus; it was the strangest thing ever! I couldn't believe what I was feeling, and it hurt, but I didn't care; I was hungry, and it tasted so good.

After almost finishing the meal, the pain from swallowing mostly passed, so I hopped in the shower and then passed the fuck out. They expected us to be at work the next day, so I pretty much just slept over the next eighteen or so hours.

Everything felt like it was moving at light speed during the next few days. Warrior Forge had started in late May; they handed out the sniper memorandum on June sixth, and the tryouts were from the twenty-sixth to the twenty-ninth, so I spent almost a month training the cadets. After the tryouts, it was back to the woods to help train them again.

After about a week, one of the sergeants pulled me aside around the tent area.

"Congratulations!" he said, grinning.

"What are you talking about?" I replied. I was confused; I wasn't sure what they meant.

"Oh," he said. "Well, I guess you haven't heard! You made the sniper team." He clapped me on the back. "Congratulations!"

I sat in my cot in the woods that night, thinking about what I had done. I was in shock; I had made it happen. I was more shocked that I achieved a lifelong dream simply through hard work and preparation. From then on, I felt like I could do anything, and all my insecurities went out the window for the most part. I grew up wanting this, and I had achieved it. Shy Boy from the past, the person who wouldn't even smile at a grocery store customer, seemed gone entirely. Now, I walked everywhere with my head held high, reveling in my victory.

I had to pack up and get ready to return to the company; I was relieved from Warrior Forge because making the sniper team put me into another platoon with Clint and Stephen. As predicted, they were the other two selected, and we all came from the Dirty Pirates. This indicated that 3rd platoon was the best platoon in the best company in the best battalion in that entire Brigade, and the Alpha Company sniper team comprised its best men that wanted to step up to the task. There were other outstanding soldiers, but they chose not to try out for some reason. I never understood it: why didn't everyone want to try out for one of the spots? Now, I think it must be self-doubt. It's a bitch; we all get into our own minds.

What I experienced with this sniper tryout was one of the first and most powerful examples of setting my mind to something and accomplishing it I had ever experienced. I think my mindset must have bloomed from the days when I dedicated my entire body and soul to Plant ID with Sly. I knew even then that if I stayed true to my commitments, put my head down, and just did it, the reward would be dream-fulfilling.

Sergeant Eddie talked to me before leaving the squad, telling me he was glad I got picked for this team. He was going to miss my contribution to the squad. I realized I didn't want to leave when he said that. I loved everyone I worked alongside. However, I was looking forward to the training and position I was going to.

We were now part of 4th platoon; a weapons platoon made up of tankers, mortars, and snipers. The platoon sergeant was a tanker, and he was a good sergeant. However, he wasn't exempt from being an asshole. I was glad to have him there, fighting on our behalf when everyone else played dumb.

The platoon leader was a young Lieutenant, and he was an exceptional leader as well. Though it wasn't long before he was gone and replaced by Lieutenant Appleton, I was lucky to be under his wing for the time we had him, Clint, and Stephen as my first line in the chain of command. Clint spent nine months at Ranger school recycling multiple times, so he has exceptional knowledge of the Ranger ways. I soaked up all their knowledge like a sponge.

Soon after making the sniper team, our sniper rifles started coming in, and I was excited to see what kind of hardware we would finally be sporting. I was called down into the arms room to go and see what we got in, and my mind imploded when they handed me our sniper rifle. It was a bolt action rifle in three 'o' eight caliber, the same gun I grew up shooting.

"Holy shit," I said out loud. "My old man was right!" I grew up groomed for this position, and dad was right all along for once in his life.

A Co 1ST BN 17TH IN
5TH BDE 2ND ID (SBCT)

"DOG SOLDIERS"

Certificate of Achievement

is awarded to

PFC RICARDO PEREZ

From 26 June 2007 to 29 June 2007

For completion of the "Dog Soldier" sniper assessment and selection. A grueling 65 hour event that tested Land Navigation, General Infantry knowledge, Physical Fitness, Field Craft, and Team Work. PFC RICARDO PEREZ displayed the drive, motivation and perseverance to succeed where others have failed.

CPT, IN
COMMANDING

LTC, IN
COMMANDING

Earning this showed me what we are capable of

Shortly after making the sniper team, Stephen got orders from the big Army to change duty stations. I was thankful for the time I got to spend and learn from him and didn't want him to go because I knew there was a lot more I could've learned from him too. He and a few others who got orders left in no time, and it was just Clint and I remaining on the sniper team. He was promoted to the sniper team leader, and I moved to the shooter position from the security spot. Not that the "title" of the spots matters much, as we were all expected to rotate through responsibilities. The only restriction was that the "team leader" was the only sergeant, keeping the other two positions from gaining rank.

On July 30th, 2007, Clint and I were off to the main post for a few weeks to be students in the pre-sniper course at the Brian Mack Advanced Skill Center. Leading the school was a crazy hardass with Special Forces credentials. However, our main teacher was an old school special forces guy we called Mr. E. He was a cigar-puffing hardass who was an encyclopedia of ballistics and combat information.

As we sat in our assigned seats in the first class session, two men to a desk, Mr. E walked in and talked about all kinds of things involving taking a shot. I don't remember what he said, and it sounded like he was speaking in another language. Things he showed us that day in the classroom were hard to put together on paper, but once we got to a range, things began to make a lot more sense as we employed what we went over.

Before we arrived, I expected Clint as a sniper buddy for school, but they separated the company teams. All the snipers from the brigade were at the school, but we worked within battalions with men from other companies.

My sniper buddy for the school was Dennis, and he could run like none other, which was fortunate for him and unfortunate for me. We ran together every single day except one, for four weeks. It sucked almost as much as the tryout to make the sniper team. We would run around the airfield, up the hills behind the Ranger battalion, and over the main post. I believe it would be safe to say that we ran a hard forty to sixty miles every week. My lower half would always be in so much pain from running every day, and every day I said I wanted to quit. I never really meant it, though, but I felt it. I never adequately stretched, so running seemed to be destroying my legs. But we were all in pain. I'm pretty sure they were trying to break us off to see who didn't want to be there.

There was nothing they could've done that would've got me to quit, but I remember one guy who waved the white flag and wanted out. The sergeant in charge made a mockery of him while we were in formation, going on a rant about him being a weak-hearted quitter. The rant went on longer than it should have, and afterward, the soldier was free to go back to his unit. Somehow seeing others give up made me even more motivated; I didn't want to be that kind of soldier.

In the sniper class, Dennis and I were some of the best. With him spotting and me shooting, we were consistently the team with the most bullseyes hit at the range. It was a highlight for me when they decided to hold an offhand (standing) shooting competition. At this competition, the range we shot on had metal targets that were twenty inches by forty inches, a silhouette of the average male torso, in one-hundred-meter increments going out to 500 meters. That was the KD (Known Distance) portion of the range, and on the other half, the metal targets were put out at random distances from one-hundred to 1000 meters for us to "mil and drill," quickly acquire distance with our scope, and engage.

A few of the random targets encroached the KD portion of the range, and they gave each sniper six rounds: five to shoot at each distance on the KD side and a sixth for a tie-breaker, just in case more than one sniper hit each target in succession. As confident and cocky as most people went up, they quickly learned that hitting a target through a 10x magnification scope while standing is no easy task. Most sat back down in disappointment.

Clint's sniper buddy, another tabbed and scrolled ranger like him, went up as cocky as he could, walking swag in full effect. He was physically the biggest guy at the school. He walked up, ready to engage the first one-hundred-meter target, pulled the gun up, aimed, and fired.

"MISS!" The cadre yelled.

Everyone there started laughing; even the shooter, as he walked away from the target, laughed with us.

"Yeah, I thought I was ready for that one," he chuckled.

When it was my turn, I employed everything they had been teaching me. Holding the rifle in my right hand, I stuck my left arm through the sling, wrapped it around my arm, then bent my wrist to make my left hand flat for a shooting platform. Then I cocked my hip out, put my left elbow right above my left hip in the soft spot, and looked through the scope. The crosshairs were out of control!

I tried to calm my breathing by taking a deep breath and focusing on the path the crosshairs were taking. The crosshairs will never sit still when you engage targets standing unsupported with a magnified weapon. It's about controlling your breathing to allow a pattern or a natural rhythm to take over, so the crosshairs move in a controlled way.

Once you see the pattern, you pull the trigger right as the crosshairs are about to move onto the target, and the delayed reaction between the time it takes your brain to send the signal to your finger to pull will be just enough for the round to land on the target.

After a few seconds of allowing myself into a natural rhythm, I engaged the closest target one hundred meters away. It was a reasonably easy engagement, but the difficulty increased exponentially with every one hundred meter increase in target distance. After hitting the first two targets with relative ease, my confidence only got bigger while my targets got smaller.

I continued to employ this technique, and it got me to the finals with only one other sniper from a different battalion.

In the finals, I went first, probably because my opponent had gone first to shoot the initial five targets, and this time, our target was out around 550 meters. I adjusted my scope, repeated the steps, and fired. It felt good leaving the barrel.

"HIT!" The cadre exclaimed.

Man, that was a load off my shoulders; now, all the pressure passed on to the other sniper. Employing the same techniques, he took his shot.

"HIT!" The cadre announced again.

They handed each of us another round to engage again, but he was to go first this time. I anxiously watched and waited, and when the rifle finally went off, the cadre yelled, "MISS!"

Instantly, all the pressure left me. I kept repeating to myself, *if I hit it, I win; if I miss, we tie.* I knew I had nothing to lose.

This time, I took my time and ensured I got control of the movement pattern. The "infinity" pattern that my breathing created with the crosshairs brought the right loop over the target's right shoulder and down across its torso, swinging left and up, looping back right towards the torso and repeating. I decided to pull the trigger as the crosshairs came from right to left. I pulled the trigger as the crosshairs started to approach the right shoulder.

When the round left my barrel, I felt as if I had pulled the trigger a split second too early, as if I were a professional basketball player who could tell when their shot had too much or too little behind it. I didn't hear the ring of feedback from the round impacting the metal target, so I thought I missed it. I instantly slumped my body language, expecting to hear the cadre's yell of my failure.

But then, the cadre suddenly yelled, "HIT!"

I couldn't believe it. After the cadre said I hit the target, I turned to him.

"I hit it?" I asked, puzzled.

"Yeah, you just barely caught it on the right shoulder area," He explained.

When he said that, I was elated. I had pulled the trigger just early enough to catch the target barely, and his confirmation of the target's side that I hit reassured me. This was how we tested people who wanted to be on the sniper team after Stephen left: We'd take them to a range and see if they could tell from which side we were missing a target by being the spotter. Spotting is the most challenging job of a sniper's responsibilities, and technically it's the sergeant's position on the team.

We left the range shortly after the competition. Before leaving, one of the sergeants took my name and other information, including my smoking habits, which I found a bit strange. I didn't overthink them taking my information, but documenting my achievement for the day is something I'm very grateful for.

BRIAN MACK ADVANCED SKILLS CENTER

PRE-SNIPER COURSE

Daily Evaluation Sheet

Name Perez (last) Ricardo (first) (mi) SSN

Rank PFC Unit 1-17 Ac Date 8-7-7

Motivation/Attitude:

Class Participation/
Good/Poor Comments:

Test Scores:

Negative/Positive Remarks:
Fantastic shooting 2 out of 2 shots
Center mass off hand
at 500yds

I almost fell over when I found proof of that legendary shootout.

On the way back, I reveled in the feeling that I was gifted with the bolt action gun; it was almost as if it was an extension of me at that point. One day, when we were engaging moving targets that move at a walking speed, the walker decided to become a runner, reducing the target window to fractions of a second. Instinctively, I snapped the gun right to follow the target, pulled the trigger, and hit Clint's hat, which he put on the cardboard silhouette's head.

The moment I shot, I wasn't sure if I had hit the target when they did that, but when we changed spots from shooting to moving targets in the pit for them, they confirmed that I hit the target when they ran it.

We continued in training, and I made one major mistake when using the mil dots in the scope for range estimation and didn't realize it until the last few days. I did well overall and graduated from the school with an overall score of 84.57 percent, which means shit as I don't know what they used to calculate that number. I imagine it's not the same as the grading system in school.

I was happy I didn't have to do daily death runs anymore on graduation day. This was a reasonably grueling course, and I had finished it. That was a proud moment and much more enjoyable than the zombie walk I gave at my high school graduation. Here, they'd taught us a lot, and I cherished every moment of it, even the running. I felt as if this experience brought the beast out in me.

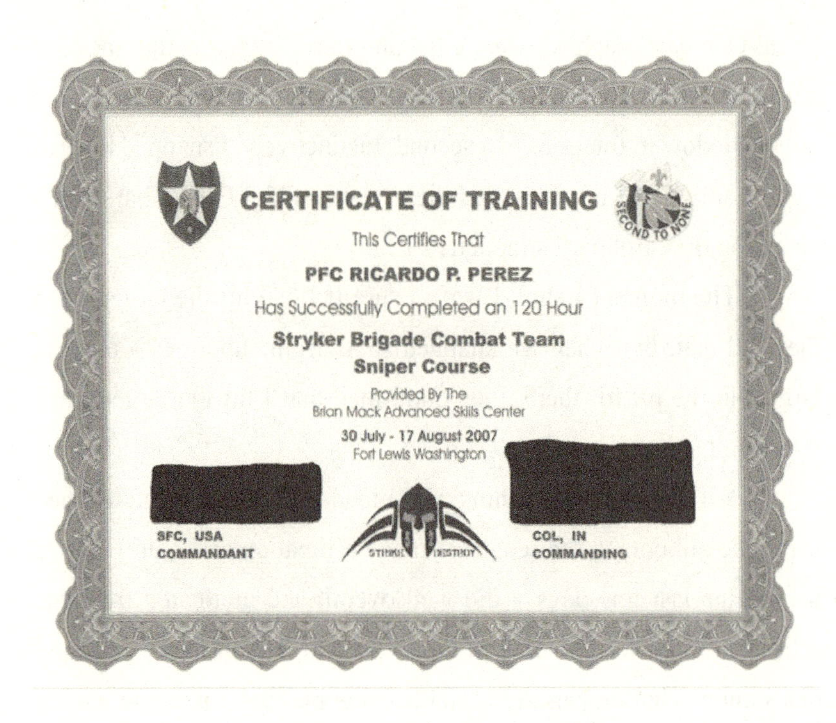
CERTIFICATE OF TRAINING

This Certifies That

PFC RICARDO P. PEREZ

Has Successfully Completed an 120 Hour

Stryker Brigade Combat Team

Sniper Course

Provided By The

Brian Mack Advanced Skills Center

30 July - 17 August 2007

Fort Lewis Washington

SFC, USA
COMMANDANT

COL, IN
COMMANDING

A lot of pain and learning to earn this

CHAPTER FOURTEEN

ONE OF A FEW

I never thought I would be a good runner, having never done it at all in any actual capacity before the military. But after all the running at Pre-Sniper training, I was going to smash the next run on the PT test, and I did just that.

To ensure I would run a new personal best, I did the run with a Camelbak Hydration Pack with just a few ounces of water to minimize the weight. I did this because I was going to run myself so hard during the two-mile sprint, and having a bit to drink (but not too much) would help me just that little more.

The Commanding Officer of our Company and First Sergeant Cummings gave me a bit of a hard time about this choice, joking that they weren't sure if it was against the PT testing policy. They let me keep it on, and I gave the fucking run everything I had, and in the end, I was on the verge of puking, and if I hadn't had the water to sip on, I would have. Suddenly, having a hydration source wasn't such a jackass move anymore. I was even asked for water by many people after the run. I didn't have much to spare, but I saved a bit, and after taking one last sip, I let them pass it around.

This was the second time I noticed authentic results from my hard work and dedication: I succeeded in what I set out to do. Most of my PT runs timed in around the early fourteen-minute mark. I had clocked in at twelve minutes and thirty seconds for the official run, beating the hundred-point scoring maximum by thirty seconds. In the end, the run helped me score over 300 points total for all events, maxing out my PT score for the first time. I was happy to have accomplished that, and the run would have been the only way I'd be able to do it.

After completing the pre-sniper course, it was back to Expert Infantry Badge, or EIB training in preparation for the EIB testing days, from the 22nd of October to the 8th of November, 2007. EIB tests an infantryman on everything he should know and be physically capable of doing. The test includes everything from throwing hand grenades (one of the lanes that fail people the most) to putting your gas mask on correctly, which includes clearing and sealing it in less than nine seconds, which was also a bit challenging.

EIB training consists of around forty to fifty different tasks and events. Those who pass all events in their first try are distinguished by what is known as "True Blue," a light blue hue is the color of the Infantry. Nothing distinguishes True Blue from those who tried a second time and still earned theirs with marks on their card. Still, the battalion commander Morton held a formation to honor all of us who earned the True Blue distinction by giving us a battalion coin. It was the second coin he had given me, following the coin I got when I was selected for the sniper team.

I almost didn't earn the True Blue distinction; the hand grenade lane almost got me. Even though I had done it flawlessly multiple times in practice, I was short-handing them on the day of testing. They gave us three training hand grenades, and the first two I threw didn't feel good, likely falling short of the trench I was supposed to get them into. I took the kneeling position behind the chest-high cover with my last grenade and True Blue on the line. Then, in one flowing motion, I threw the grenade and hit the prone lying on my chest. This time, it felt good leaving my hand. I remember it feeling a bit better than the previous two, but not good enough for me to be confident that it went in. As I lay there wondering, I shifted over to my shoulder to look at the sergeant in charge of the lane.

"It got in?" I asked, excited.

"Barely," He said with a smirk. "You got lucky."

I was thankful I got the last one and moved on to the next station.

Things slowed down a lot after EIB because we were rolling into the holiday season, and things with Eve and I were wearing thin. In October, she had flown back to Texas for her mother's graduation and decided she wanted to stay there until I came in December.

I drank alone during those weeknights while she was gone, since they weren't PT-ing us to death with the holidays coming up. I mustered up the courage to message my ex-crush, Celina, over MySpace during one of those drunken nights. I was just that damn lonely.

I didn't think she'd reply at all since she had turned into a bombshell cheerleader at a local university and was entering her junior year. But she did. Somehow, she remembered me.

Every evening after work, I either messaging or talking on the phone with her. We spent a lot of time on messenger, and we would occasionally use it to video chat. She was all I could think about, and I was scared to be frank with her about me and Eve's struggling situation.

Eventually, though, after a month or so, I told Celina that Eve and I were unhappily married, hoping she would understand. I explained that things were mutually disintegrating and that we likely wouldn't last much longer. I said our marriage had already gone on way longer than initially planned, and I lived in an unhealthy fantasy world.

During Christmas leave in 2007, I flew down and spent almost the entire break with Celina. I saw Eve for a few hours and a day with my family, but I spent 90 percent of the time with Celina. It was a great two weeks, and the time went so fast I didn't want to report back.

On the last day of leave, I visited Eve to tell her that I thought we should end our relationship, and she agreed. It was as easy as that.

It seemed like it would be as easy as we had intended, but instead, her mother decided to try and drag things on. She was coercing Eve to try and get something out of the divorce, which was a dirty move. Thankfully, Eve eventually signed the paperwork, and our relationship was done, with no strings attached.

*****

I flew back to Fort Lewis and immediately began preparing for my flight out to Fort Benning, Georgia, but I would attend the U.S. Army's Sniper School this time. Sniper school has a 50 percent attrition rate and was five weeks long at the time. They've since added a few weeks as they were pressed for time and were soon going to be adding the Army's newest sniper weapon system, the M110 SASS, to the curriculum.

Sniper school was mostly mentally tough; they didn't have time to try and physically break us; we only did PT for the first few mornings. We spent the rest of the time in the classroom, outside stalking without getting spotted, learning to track via footprints, and performing other exercises like target detection. There is a big misconception that the majority of what snipers do is shoot. In reality, shooting makes up about 25 percent of a sniper's overall time in school.

Clint and I were still sniper buddies. While we were in Georgia for the course, he went through it with a hernia that he got surgery for shortly after we graduated. We butted heads one morning during PT because I was doing the exercises faster than him.

"Hey man," he said in an offended tone. "You're not even doing the full sets."

"You have no proof of that," I retorted.

We immediately stopped working out and got in each other's faces, angry from the adrenaline we'd built up from the exercise. I put my arms behind my back as I usually would've if we weren't at the school; even though I didn't have to, as students, there was no rank distinguishing us at sniper school.

A highlight from this time was probably passing Stalk Week, which is an entire week dedicated to "hunting" the school cadre. The challenge is so complex that it's the primary way they fail people out of the school. The cadre always ran the same lanes every month and knew where all the avenues of approach were, and it was tough to get in position with a seasoned sniper looking for any movement. It was the third stalk that I passed, and I was close to being caught after taking my shot. Once I was in position, I fired, and eventually, the cadre I engaged walked another one of the cadres close to me and gave him directions over the radio. He walked within eighteen inches of my barrel but thankfully was given directions away from me. Finally, one of the Rangers from Ranger Battalion had to finish to graduate on the last stalk. Lucky for him, they put his buddies in charge of keeping time, so it didn't matter that he came in late. It bothered me to see him get special treatment because I knew that I wouldn't have been so fortunate if I were in that situation.

We had weekends off during the entire five weeks, and thankfully no one went out and got in trouble over the weekends. If news of trouble got back to the cadre, we wouldn't have been able to keep having those weekends off.

My worst school experience came on one of these weekends, which involved a lot of booze. I went to some guy's barracks—these barracks were familiar because we would often hang out during airborne school. I was able to find one of the guys I used to know, and he told me that the other regulars had been in a lot of trouble recently. Some of them were on the verge of getting kicked out, so I asked how he was doing and was thankful for the update. He was doing better than I suspected.

I asked about him because this group of guys was some of the most suicidal soldiers I had come across, other than the military cooks who worked to support RIP and the Ranger Battalion. The Army shafted them because they originally signed up as Reserve or National Guard, and when they changed to active duty, the Army gave a bunch of them orders to support Drill Sergeants at lanes on Fort Benning to help train the basic trainees. My buddy then told me he fucked up by re-enlisting to change jobs and become a helicopter mechanic but got a DUI shortly after, so they took away his opportunity for that training and kept him there helping the Drills. I hurt for him because I remembered from back home just how easily one could get into trouble before realizing it.

Ironically, that night, we got absolutely hammered. We drank a lot of hard alcohol, played beer pong, and drank pretty much everything in the house. I vaguely remember walking back to the liquor store, drunk with a few others, to buy more cheap vodka. When we got back, we couldn't drink it because it smelled like rubbing alcohol.

That was the first time in my life I had ever completely blacked out from alcohol, and I had been on the phone with Celina at some point during the blackout. The next day I called her, expecting things to be fine, but they weren't; apparently, I had said all kinds of stupid, mean shit while drunk. I couldn't believe it, but she said I did, and I apologized, hoping she wouldn't break up with me. Thankfully, she forgave me, but I sure as hell didn't learn my lesson and continued to party like there was no tomorrow.

As sniper school graduation approached, the previous graduating classes decorated cardboard silhouettes like those we shot. Still, this time the Rangers in the group thought it would be funny to buy a cardboard cutout of Hannah Montana, so that's what they did. We decorated her proudly, and she hung from the rafters for who knows how long. I know you can catch a glimpse of her in the Army Sniper School documentary shot in 2008.

CHAPTER FIFTEEN

THE VEIL IS PEELED BACK

Clint and I made our way back to Fort Lewis as fully-qualified snipers, and we were starting to discuss bringing in another person to fill that security spot on the sniper team. I pulled for my brother Duncan to be the one; I had spent enough time with him on the weapons squad to know that he was a good candidate for the position. As much as we wanted to have input in the choice, they took it out of our hands; and went with the person who scored the highest PT test in the company. His name was Timo, and though he scored the highest, he surely wasn't very athletic. He lucked out by being scored by a sergeant from his platoon during the PT test.

While going through all this consistent training over so many months, we got a few new guys in our company, and there was one guy named Gilbert, who was a particular case. I heard many horror stories of things that his squad leader, Randall, made him do, but being in a different platoon, I usually only witnessed the things he endured around the company. I won't rehash the allegations because they aren't confirmed, but I feel they likely transpired in my gut. This belief came about because of what I *did* witness with my own eyes: the constant, daily hazing. Randall hazed the piss out of him because he didn't have the mental capacity to do any basic infantry tasks. There was only one time Gilbert, and

I ever talked man-to-man. It was my buddy Paulo's birthday, and I wanted to wish him a good one.

"Hey," I said, coming over to Paulo while he sat on the ground on the first floor of our building.

"What's up, P? Happy birthday, man." I said as I began to sit next to him.

"Thanks bro," Paulo replied. "What you up…"

Before he could finish, Gilbert jumped in. "Hey P," he said. "How old you gonna be?"

"Twenty-nine."

"Twenty-nine, huh?" Gilbert said. "Uh, I think you mean ninety-two!"

Confused, P and I looked at each other, dumbfounded, and then he continued.

"You know, like twenty-nine backward. You know, I'm going to be twenty-three in January."

"Wow," Paulo said, trying not to laugh. "Wow man, you're like a mathematician or something."

"Nah, I'm not. But my wife is."

I can't remember where the conversation went from there; I think he may have just continued on his way, but that was the only time I ever really heard him converse with anyone. I don't even think I participated in the discussion; I just sat back in wonder and sadness at what he was going through daily.

I realize the frustration a sergeant might have with a soldier who has no "soldiering" ability, but to purposely make his life hell because of his mental capacity was just wrong. Physically, the worst the hazing got for him was having to carry around a five-gallon water jug everywhere he went. Often, Randall decided he felt like making him hold it over his head. This was just one of the various psychological torture methods that Randall employed. Still, one that even the sergeant-major took part in was the time a Rottweiler dry-humped him while playing fetch within their vicinity. Gilbert and another soldier were forced to low-crawl on their bellies across our battalion's field area. Everyone wanted to see it happen, myself included, mainly because it was harmless, and I was sure it wouldn't happen. Then, almost as by the company's collective will, the dog walked over and tried to relieve himself of some of his genetic material on their backs while they crawled. That was when the entire company laughed, and while we did, sergeant major took his phone out to photograph or record the incident.

If the other allegations are real, I wish I were there when they occurred to document them here. I may have likely been in the vicinity when these things took place, but I never directly witnessed them. The hazing got so bad that after what seemed like months of him going through it all, one day, Gilbert was suddenly gone from the company. The word was that supposedly another sergeant finally pulled him to the side, wrote down what he needed to tell the investigating body, the Inspector General "IG," and had him read it to them to help him out with the hazing.

The investigation had started before I went off to sniper school, and I thought, *Awesome! There is no way that Randall is NOT going to get in trouble; everyone saw everything he did daily!* Well, the days went by, and Randall was still Randallding around the company with no worry in the world. In my mind, the whole investigation was a done deal, and I wasn't expecting to see Randall or Gilbert at the company upon my return from Sniper School. But, as luck would have it, it turned out that they kicked Gilbert out of the company, and Randall moved from 3rd platoon to 4th platoon, which was MY PLATOON! I couldn't fucking believe it!

After all the stories I heard and the things I witnessed, the investigators did absolutely nothing to change Randall's behavior. I'll never forget one of the first times I had to deal with Randall firsthand. Most of the platoon was in the upstairs conference room, minus the sergeants. We were there bullshitting (which was the theme for the next few years leading up to deployment for the most part) and talking about all kinds of random stupid shit more than likely. Then, Randall walked in. The room quieted down, not knowing if he had something relevant to tell us or if he was just there to bother us. He sat down and just said that we were to wait for the word to come down on being released, and we continued to converse. We were talking amongst each other when suddenly Randall chimed in.

"Hey Perez," he yelled across the table. "You think you're a badass because you graduated sniper school?"

"Negative sergeant," I said. I wanted to take on a respectful tone, even though I had zero respect for the guy. "I don't think I'm a badass for graduating sniper school, just fortunate."

"Well I carried the M14 in Iraq and …"

"Oh!" I interrupted. "So you graduated sniper school?"

He said no, then rambled on about something, and I ignored him. I should've seen the inaction by IG as a prophetic vision of the future, where no one's issues were going to be remedied and likely only worsened. As for Gilbert, I thought they kicked him out of the Army. He was every sergeant's favorite example of "The recruiter forging paperwork," meaning he likely wouldn't have even been able to get into the Army without extensive "help" from his hometown military recruiter. However, while driving around near where we worked, I finally saw Gilbert.

He was on a riding mower, mowing the grass at the Brigade area.

Wow, I thought. *Rather than kick him out the Army, they made him the Brigade Janitor!* The whole situation was, as they say, "ate the fuck up."

*

Summer was slow; I participated in Warrior Forge again, but with 4th platoon instead of 3rd. The entire time, it sucked: we stayed out there days at a time and didn't have nearly as cool a job as we had the year prior. We rode around on the 6x6 utility vehicles and took water places and other monotonous tasks.

I had a close call that summer when I was driving the FMTV. It was a large vehicle that we used to transport troops and equipment, and one day I was behind the wheel. After bringing it on post to refuel, I drove it back out to the training area. At the end of a long road, I finally came to the stop sign right before a two-lane highway that I had to quickly dart across to continue on the back roads to the training area. The guy riding with me wasn't paying attention in the passenger seat, and I looked left, then right. That was when I saw an old, small, white truck far down the road. I didn't ask him to clear me because I saw how far away the truck was. I quickly looked left again and then straight to continue on my way, knowing I had enough time.

As I pulled the vehicle about halfway across the road, a white flash appeared and disappeared just as fast. I looked at the other soldier, and, as expected, we both freaked the fuck out! Looking to the left, I saw a white truck had flown by and went along as nothing had just happened.

We came within inches of a wreck more than likely, and I was shook. I began apologizing but explained that the truck was a reasonable distance away when we stopped at the sign. It was just going way faster than the speed limit. The driver of that truck didn't care. If he cared that I was halfway in the road, he would've honked his horn or more than likely pulled over to see what unit the vehicle belonged to.

For the rest of the summer, we did a lot of sitting around. This didn't change much, either, when Warrior Forge was over. We rarely trained. Instead, we just sat around at the company most days and learned how to send a 9-line medevac to call for aid in a firefight and other basic infantry stuff. Not that it isn't essential to learn those things, but it got to the point where I knew everything there was to know and wasn't allowed to participate in Q&A time anymore.

I was dying inside for some action, and in Fall 2008, our brigade command selected our battalion to train with the Japanese Self-Defense Forces. That solidified something to me: our battalion was considered the best conventional unit battalion on post at the time. Why would they choose us if we weren't? At a minimum, we were the best battalion in the brigade because if we weren't, they would've had one of the sister brigades take the helm. But this also scared me because it meant if we were the best, then how bad was the worst?

1st Battalion, 17th Infantry

CERTIFICATE OF ACHIEVEMENT

IS AWARDED TO:

SPECIALIST PEREZ RICARDO

FOR OUTSTANDING PARTICIPATION IN THE RISING WARRIOR EXERCISE. YOUR DEDICATION TO DUTY AND MOTIVATION BROUGHT GREAT CREDIT UPON YOU, THE 22ND INFANTRY, AND THE 1STBATTALION, 17TH INFANTRY REGIMENT.

SIGNED THIS 19TH DAY OF OCTOBER IN THE YEAR 2008

CSM, USA
COMMAND SERGEANT MAJOR

"BUFFALOES"

LTC, IN
COMMANDING

It was an honor to participate in the Rising Warrior Exercise

During our big Army annual required morale evaluations, we were encouraged to write down how we felt about the unit, so all of us did. They told us to do it to help bring change to the company, and year after year, we did, but nothing ever changed. What I've since concluded was that the grit and hard work of the more competent lower enlisted, along with a handful of competent sergeants, was what hid the incompetence at the higher levels.

CHAPTER SIXTEEN

COMPLACENCY KILLS

The next step in our training was when a battalion of Japanese forces came to train with us. Since I was a sniper, my sniper buddies and I trained with our Japanese counterparts. We were excited to have something to do, and over the few weeks they were there, we had great fun.

I gave a tutorial on how to paint the rifles properly but helped with many other things as well, of course. We participated in some of these classes but had them go through a shortened version of our sniper school. We even used range 21, the range that no one ever used. That range was something that I was fortunate to have experienced. We never used that range as it wasn't practical for us to train on. We were better off on our standard range, range 21, because of how it was laid out. Range 19 was likely only used occasionally by the different special operations forces on post, and every time we drove by it, I would try to peek through the woods to see if I could see if anyone was using it. It normally looked like the small ghost town it was meant to mimic.

After the Japanese forces left, it was back to sitting on our thumbs. There wasn't a great command chain that motivated us, resulting in everyone's boredom and depression. We occasionally went to the range, mostly when the required annual rifle qualification was due. I think that only happened two or three times after the pre-sniper course as far as brigade-level sniper training. As time went on, I started seeing just how down in the dumps everyone was. This, to me, was an alarm signal: the brigade was likely a ticking time bomb, and I needed to get out of there as soon as I could.

My first course of action was to drop my warrant officer packet. A warrant officer packet was the easiest way to get out because all it required was paperwork being submitted to join a pool of applicants for warrant officer school.

I researched, downloaded the form, filled it out, and talked with others about it. I was disappointed to find out the packet required a signature from my battalion commander. But, as someone advised me, "There is no way in hell he is going to sign away one of his battalion snipers."

I went back to the drawing board and found a program called the Bonus Extension and Retraining (BEAR) Program. This program's purpose was to recruit people from overstrength jobs into desperately needed roles, like special forces or counterintelligence. I decided my best course of action was to go and talk to the chaplain, as he surely wouldn't have any bias about me leaving the unit.

I walked into his office and explained the program to him and how I wanted help. The chaplain sat back in his chair and listened intently.

"Look, Perez," he said when I finished speaking. "For those jobs, they like soldiers to have deployment experience. Just deploy with us, and you can do that after we get back."

I'm sure the shock on my face said it all; I don't even remember what I said after that. I likely just got up and walked out. At that point, I realized there was no getting out of the unit without going and trying out for Special Forces, just as dozens of men from the battalion had done before me. Other men were so desperately using this course of action to get out of the unit; my sniper school slot was threatened by Cummings when I was even around people talking about it. That was going to be my only ticket out, and I had zero doubt in my mind that I would give my all, get selected, and reach salvation from the hell I was experiencing.

Allegedly, going to SFAS was also a question at the sergeant promotion board. They had no intention of promoting anyone who had intentions of trying to keep their sanity or better themselves.

My platoonmate Axel asked to get schooling because we weren't doing anything and he wanted to increase his ASVAB score to maybe reenlist for a better job. He was able to start classes, which was a truly historic feat. However, having him go to the main post a few times a week was just too much to ask. He needed to keep in close contact with the company, so it seemed that no one felt obligated to help him with his personal and professional development.

A few other guys from the platoon tried to develop something to make themselves feel better, and I heard it as I walked into the upstairs conference room. As usual, one of us started bitching about how shitty everything was when someone else piped up.

"Hey guys, lay off," that one person would say. "It could be a lot worse."

When I walked into the room, at least four or five guys were yelling in unison, "It could be worse! It could be worse!" At that point, a few others joined in, actually chanting the new mantra. I asked what was going on, and they said it was their recent play trying to make things change.

I responded, "Hey guys, lay off. It could be a hell of a lot better."

After that, they gave me shit, and we all had a good laugh. That systematically killed the positive rendition in its tracks. I've always considered myself a realist, so I wasn't going to let something so simple brainwash my brothers into thinking that we could have been better off.

I sat up in my bed that night, and all I could think was, *man, this is the bottom of the barrel in terms of military chain of command. And if this place is this corrupt and fucked up, I can't imagine the level of corruption and fuckery that goes on at the presidential level.*

All the Hollywood movies we've all seen are likely just bedtime stories compared to what goes on behind the scenes in the real world, the place with real consequences. Out of options, Tracey, Terrence, and I signed up for the same Special Forces Assessment and Selection (SFAS) course. They were both mortarmen within our platoon, some of the smartest within the company, and all of us were fairly close. When our orders came down from the special operations command, our company command raised hell. Tracey was fortunate to dodge a lot of the bullshit as he was attending language school at this time. A few guys left our company in the years prior after being selected, but there were many more from other companies. A mortarman we knew from Charlie Company had gone three times before he was selected. He figured perpetual hell at SFAS with at least a chance of things getting better was a better alternative to the daily shitfest we endured back at 5/2. They had Terrence and me standing by their offices as if we were in trouble.

"You causing trouble?!" The other guys would ask, walking by.

"No," I replied. "We're just going to SFAS."

They'd usually say "oh" and walk off because it was no big deal, but "the leaders" were treating it as if we got a DUI or failed a drug test. Finally, my sniper team leader, Jack, came by.

"You just burnt a major bridge," he told me with a smirk. "Better not fail."

"Roger, Sergeant," I said.

Many people were confident that I would be selected and were mad that I would likely not deploy with them. The one who was most notably angry, First Sergeant Cummings, told me I would probably be selected. He was selected in his career but chose to stay in the conventional army to make rank.

The fact that Cummings was mad at me for signing up hurt me and made me sad to be leaving the guys I loved so much and trained hard with every day. At the same time, though, it bothered me that I had to live my life in such a shitty place where everyone was depressed daily. I desperately wanted everyone to go try out. It was worth it to at least attempt to leave that hellhole. It was the only way we were ever going to be able to leave that place; we just waited until it was too late.

This was the moment I realized I had become complacent. This was strange since I had worked so hard to make the sniper team. I once took pride and honor in the position I held, but slowly, they chipped away at my morale until I couldn't take it anymore.

The Charlie Company sniper team leader went to that class with us, and he was also someone everyone knew would get selected, but who also decided to deploy as the sniper team leader for his company. The biggest thing that made everyone a terrible sport about this whole thing was that we were going to SFAS right before NTC, and we didn't tell them we were going because they had already told us not to go until after NTC. We did it because we had been working out and getting in shape for our journey into the two weeks of hell, and if we went to NTC first, we knew we wouldn't be able to work out often and wouldn't be prepared for SFAS. As soon as 1st Sergeant Albie got wind that I was going to SFAS, he called me into his office.

"So Perez," He said. "I hear you're going to SFAS?"

"Roger, First Sergeant."

"You're going to fuck your team, huh?"

"First Sergeant, we were a two-man sniper team for over a year and operated just fine."

"Alright, then."

It was a massive debacle for us going to SFAS. They threatened us with Article 15s (the army's punishment, i.e., extra duty, reduced rank, reduced pay) and separation of the Army for disobeying them. They told us not to talk to the SF recruiters, which we knew wasn't a lawful order. Most people got a kick out of our interest in SFAS because it was dumb that we were being criminalized for doing it when other people in the company were shitting on cars and stealing snowboards, all the while avoiding the sort of punishments we got.

Knowing that they were going to bitch and whine and could do absolutely nothing about it was highly satisfying. We ended up getting negative counseling statements, which were just a slap on the wrist. I felt terrible because the worst part about it was that the NCOs in the platoon blamed the other soldiers who weren't going because they knew that we were going. Terrence and I continued standing at parade rest outside of their office. That didn't last long, though, and they eventually released us for a formation outside.

They got us together and looked straight into our faces.

"You guys are the problem," they said as they deflected blame to everyone else in the platoon. "You didn't try to stop them from going and talking to the recruiters."

Fuck that. I thought. This whole situation was dumb because the NCOs told us not to talk to them, and we did it anyway.

In any case, we went off to SFAS at Fort Bragg in late November or early December, and it was cold as hell. It snowed so much that non-essential personnel was required to stay home.

I'm not going to go through the details of what happens at SFAS in this book; if you want to go and watch everything that happened, you can watch the documentary "fourteen days in satans home." When I was there, they were there filming what was to be the last two-week SFAS course, as it's now a twenty-four-day-long course. They added nothing new to the program; they let the candidates sleep more during the selection process.

I was doing well in the course until the log physical training session. They have a bunch of candidates on a log, doing exercises with it, and then they have us "Roll!" When the cadre yells, we are supposed to lay on the ground and roll as fast as possible. Rolling throws the equilibrium off and causes most people to start throwing up. I knew we would start with the log and rifle physical training that day and decided not to eat breakfast since I knew I would throw up during the log session. The idea behind not eating my meal was to keep myself from puking the food I ate an hour-and-a-half prior.

Not eating proved to be a fatal mistake. When in the log exercise, I still had to puke, but instead of having something to throw up, I was bringing up pure bile. It was the rawest, painful, most dizzying puking I've ever experienced and will likely ever experience. It hurt so badly I was hammer fisting the ground while my stomach wrenched in an attempt to get something out, and what came up had tints of green and yellow. My esophagus was on fire as I could feel the acid destroy the lining in my throat.

But as men dropped out, I continued. There was nothing that would stop me from earning that green beret. I've never watched the documentary, as I lived most of it, but I can almost guarantee you that there is no footage of me puking during those sessions. It was way too violent and likely not to make the cut in the final televised version.

After the bile session, they allowed us to eat and prepare for our next task, a march which was an undisclosed distance to our land navigation training site (I'd say it was a twelve-mile hike). As everyone began filling their stomachs again after dumping them during log PT, I sat and joined them. I tried to eat whatever packaged meal they handed me but threw up right away. I moved down to the next food, trying to get my body to take it, but it refused.

"I need to eat something," I told myself aloud. "I don't know if I'm going to survive this march if I don't."

I tried and tried to put food to my mouth. The last option was my pack of raisins, and after trying to eat even just one and almost throwing up, I gave up. Even keeping water down was a task. I would have to do this hike with 50+ pounds of gear with no fuel, but I figured that I could continue as long as I didn't pass out or quit. I knew it would be a red spot on my assessment, but I wasn't going to quit. Off I went on the march, and I went slow because I had to try and make it to the end. Slow as a turtle, I went, and it was tough, but I eventually made it to the end. I wasn't even one of the last ones to the goal line, which made me feel a bit better about my slow pace. It was dark as we got into camp, and the first thing I did was go to the sick call tent to see if they would give me an IV, though I knew it was a long shot.

"What do you need?" asked the medic on duty. He looked up from his work, and I immediately realized he was annoyed.

"Can I get an IV?"

"No," he said, looking back down.

"Okay, how about some aspirin?"

"No."

I knew that getting something to help the pain wouldn't happen, but I had at least saved the best request for last. "Then, can I at least get an antacid?" I asked.

"Oh, sure," he said.

Instantly in my mind, I envisioned the waves of almost immediate relief. I was expecting the pill to be chewable. Still, nothing was going to be easy at selection. They handed me pills and sent me on my way, and I immediately took them, hoping they would help soon, but the relief was still hours away.

After the march it was time for dinner, and they made us eat, but I don't remember if I tried to eat anything after taking those pills. What I do remember from that night was it being cold while we all looked for spots on the ground to place our sleeping bags. At first, I thought I had found a good location. Leaves were blanketing the ground, and I decided to move them to lay my bag directly on the floor. That was a massive mistake because while clearing out the leaves, hundreds, if not thousands, of spiders came crawling out from the area I disturbed. It was pretty surreal, so I got my sleeping bag and stuff out of that area as fast as I could and just found a new place to bed down. It was still on that blanket of leaves, but I just laid on it this time.

I woke up the following day feeling better from the antacid doing its thing overnight, and over the next few days, while learning land navigation, I felt almost fully recovered.

They taught us the basic land navigation, and we moved on from there as quickly as we came. Many "candidates" weren't from combat jobs, so they didn't practice these skills regularly. After a few days of learning, they sent us out to try and find our designated points for record.

The night before starting the land navigation course, I felt phenomenal! I felt like I got my second wind, and before bedding down that night, they called for three candidates to come up. They called my number; I ran as fast as possible and was the first one of the three to get to them. They handed us an undershirt, a typical moisture-wicking one, but this one had an electric box attached to it, big enough to fit into the shoulder pockets of our uniform.

"This shirt monitors your heart rate and body temperature," the guy said. "Don't worry. The cadre won't use it in your assessment."

Okay, cool, I thought. *They just want some data for purposes of making SFAS safer.* I thought this because they just had a man die in the class before ours during the land navigation course. They emphasized that we were to keep our reflective vest on our rucksack and our reflective belt. It had to be visible at all times. If they weren't visible, it was grounds for getting the boot from the course.

That night, even though I wasn't feeling too tired, I managed another few hours of sleep because they were giving us about five hours to rest before starting the land navigation course. Waking up ready for the course, I was off with my heavy ruck, fully rested and ready to find my points.

CHAPTER SEVENTEEN

27 MINUTES

It was still dark because it was early in the morning, and it was a night like no other. This became problematic.

We weren't allowed to walk around using our red-light filter on our headlamps during night land navigation. Night vision devices amplify residual light like from the moon or stars, but this night was overcast. It was so dark it would've been hard to see, even with the assistance of an infrared floodlight. I couldn't even see my hand in front of my face. I'd only use my red light to find my azimuth,[1] and once I'd turn it off to start walking, it was only a matter of time before I'd take a misstep. Tripping while carrying fifty pounds on my rucksack, I toppled over.

I fell about three times when I decided, "Okay. I'm not going to twist my ankle and get dropped for medical reasons, so I'm just going to move near an intersection on my map and just camp out there till first light."

I got my map and compass out, shot my azimuth, and walked. I walked right into a wall of impenetrable native brush, so I skirted along it to see if I could find a way around. I couldn't. I tried and tried, then decided that I would shoot my azimuth in straight cardinal direc-

---

[1] Azimuth: the horizontal angle or direction of a compass bearing. The compass bearing to my point.

tions and walk until I found a "cheater post," which was just a random stake with coordinates on it. So, on I went. I shot an azimuth, walked in every direction, and no matter which way I went, I ran into that same wall of brush. I didn't know how I walked into a place without finding a way out. It would have been easy to panic.

I kept my red light on at one point, desperately trying to get myself out of what seemed to be like a maze of brush. I was done with falling over, so I figured I would wait until the sun was finally up. With just about an hour or two to go, I thought I'd take the opportunity to catch some more sleep since most other people do so during this event. I wasn't tired, but I wasn't going to sit there in the dark for no reason.

I took the reflective vest off my rucksack and did everything I could to hide, even employing deceptive techniques that would've got me kicked out of sniper school. I set my alarm on my watch for an hour and tried to fall asleep. I may have caught a few winks, but when my alarm went off, the sun wasn't even bringing color to the sky. I set my alarm for a shorter time. Again, my alarm went off, and the sunlight still hadn't come, so once again, I set my alarm for the same short time.

My ears were on high alert, even as I dozed in and out of sleep. This time, as I lay on my right shoulder with my eyes closed, I heard leaves rustling as people walked up to me, and I immediately knew that it wouldn't be good. I opened my eyes to a world lit mainly by a sun that wasn't quite yet breaking the horizon. As I turned over on my back to see who had graced me with their presence, I saw the cadre, along with two men in civilian clothes, one holding a laptop and the other with a production-quality camera.

"Are you fucking kidding me?" I thought. The sergeant started laying into me, bitching at me about how dangerous it was for me to do what I did. I couldn't believe that they had used the shirt to track me down.

I blocked out whatever he was saying and gave general responses because I was still in shock as to what had just transpired. They told me the shirt was only going to track my heart rate and body temperature and that it wouldn't impact my assessment, but it was directly the cause of their finding me.

We walked through some pretty thick brush on the way out because of where I had walked a few hours prior. I couldn't believe I walked into that area at night with no effort, and even with the daylight, it was hell getting through it all.

Everything from this point on started feeling surreal. They had me get my gear, ripped off the candidate number from all my stuff, and pushed in the back of a truck before I even knew what was happening. They took me back to the staging area. On the way, many of the cadre confided in me, suggesting that I had just struck some bad luck.

"If you ever want to come back here again," one of them said. "Don't say anything about the GPS tracker."

They knew I got fucked over hard, but what they didn't realize was that this was it. Getting selected was my only chance and hope of avoiding what I knew was to come, and the opportunity to continue a military career slipped through my fingers. The hardest part about being back at the staging area was knowing that I wasn't the only one who slept during the course. I was just the only one who got caught, and it was likely because the cadre didn't even bother actively looking for anyone that night with no residual light for the night vision to amplify.

I sat there for eight hours, keeping a tiny fire alive with sticks I found by the area designated for the "failures." I sat there moping, writing apology letters to Celina. As the day went on and people came in, the frustration built inside me. The event finally finished, and I was the only one cut from the course for that event. The cadre went to grab the sergeant major and commander of the SFAS program to do their jobs for the cameras. Sergeant major chewed me out, asking me who I was trying to hide from by hiding my visibility gear.

"I was trying to hide from everyone, Sergeant Major."

"No," he said calmly, getting in my face. "You were trying to hide from my cadre!"

"Negative Sergeant Major, I was hiding from everyone; I didn't want anyone to see me."

He continued his bitching for a bit, then the commander came up. "What happened?" he asked as if he didn't know.

"I was sleeping, sir."

"It's fine; it happens all the time."

He welcomed me back to the course in the future, which was the end of the ordeal put on for the cameras. They transported me to an area where they held guys who voluntarily withdrew and the guys who got hurt. The shelters were tent-like but sturdier like a modified Morgan building. I couldn't believe what I had just gone through, and soon I was to have the craziest out-of-body experience I've ever had in my life.

CHAPTER EIGHTEEN

TAPPING INTO THE MULTIVERSE

There are no words to describe how the next few days went, but, as always, I want to do my best to convey my experience. That night, before bed, I sat around while the other people laughed and told stories or just played card games as if nothing was wrong. Looking at them, I grew furious by the minute. It was apparent they enjoyed finishing SFAS, and all I could think was, "I don't belong here." I didn't. I didn't withdraw voluntarily, and I didn't get hurt. What did happen, though, was that I got screwed.

Hearing the other guys' pleasure when they realized they were returning to their units was even more infuriating. I don't remember if I cried; I probably shed a few silent tears because I knew what was to come. That night and the next few, my dreams became my reality, and I continued in SFAS selection. They were the most vivid dreams I've ever had to this day, and everything about them felt so real. I could feel the exhaustion and the determination, the torment of team events, and the delusions of the final road march. My mind recreated sights, sounds, smells, and everything, fuelling my desire to continue on the path to salvation.

When I woke up the next day for breakfast, there was more sitting around listening to guys have a great time; nothing seemed real. I felt like I was still dreaming because I couldn't believe the events that unfolded to get me in that tent. During one of those actual days, I was hanging outside the tent, probably tired of hearing the joyous atmosphere. That's when I saw my brother Tracey running by and called him over. He asked what happened, and I gave him the quick rundown.

"Dude," he replied. "I had to wear that stupid shit too, and when I did, the camera guys randomly popped out of the woods and started asking me questions."

I was stunned. "Why the fuck didn't you tell me that?" I asked.

"They told me that it was going to monitor my heart rate and body temperature," he said. "They said it wouldn't be factored in my assessment."

Before Tracey ran off, he told me he would probably get medically dropped from the course.

"Why?" I asked. That's when he lifted his pant leg to show me his extremely swollen, colorful knee.

"Oh my god," I said. I was shocked at how swollen and bad his knee was! "What the fuck happened?"

Tracey was running back after eating dinner the night after land navigation and ran into a bike rack in a dimly lit area. As you might guess, it hurt him pretty bad. I couldn't believe it—what shitty luck we were running into! Terrence was the last of the three from the company that went together.

I continued having my strange dreams. The strangest thing about it all, though, was that the dreams were painful. It felt like my spirit was being ripped from my body, the intense situations my sleep put me through hurt like hell! I was sore for days.

Now, I feel that point in my life caused a significant crossroads to change. I felt like the entire universe was changing to preserve the balance of the multiverse theory of our multiple selves. Instead of being a special forces soldier in my "previous" universe, the universe I was in at the moment was thrust upon me. This forced the other path of wearing the green beret for a different life in an alternative reality. What was to be a life destined for a career in Special Forces became altered by a divine force, and I was tormented spiritually over four or five days as a result.

Tracey went off, and I went back to my torture, but I had gone home before Tracey came to the tents after medically withdrawing.

Everything seemed dull on the way back home; I had no life in me, and time was a malfunctioning construct. You were supposed to report back right away to your unit if you failed SFAS and inform them that you weren't selected, but there was no way in hell that I was going back to the unit any earlier than I had to. I got home, embraced Celina, cried, and apologized to her for what had happened. We spent the next few days home, depressed. I did not look forward to showing back up to the unit. After the SFAS class ended, Terrence called me up and told me he and the Charlie company sniper team leader got selected, and I congratulated him.

"Man," Terrence said. "I was done when it was time for the final road march. I went for as long as I could, started hallucinating pretty bad, and that devil on my shoulder kept telling me to take a quick break on the side of the road and that it would be okay. I resisted as long as I could but finally gave in and hit the side of the road because I couldn't go any further. Shortly after I fell asleep, I was awakened by the documentary crew guys."

"Are you serious?" I asked.

"Yeah man," he said and then started laughing. "They woke me up, and I was like, 'shit. I gotta get up and go. Who knows how long I've been down,' so I got up feeling refreshed and pushed hard to make up for the lost time. I'm running at a good pace because I'm feeling good from my rest. Then one of the crew guys asked where my weapon was. Fuck, man. I thought I was fucked. I backtracked a long way, and the camera guys helped me find it with the bright lights from their cameras."

I couldn't believe what he was telling me. I bitched at him after that, telling him how lucky he was. He confirmed it; he was fortunate that they helped him find a weapon that he alone would've likely never come across on his own.

The documentary directly led to my removal from the course and helped him on his path to getting selected. I was happy for him, as I wanted everyone to get out of that hellhole, but it was a severe kick in the nuts. I couldn't help but feel like my "Perez bad luck" was on full blast with the knob ripped off and no way to stop the machine.

I wasn't looking forward to going back to the company; the hell we all knew we had to deal with was coming soon. It was right around the holiday break time when we got back, and we went right into it. I probably dealt with a bit of hell, but the best was yet to come. That December, Celina flew down a couple of days before I would be able to because the flights were a little cheaper. I had to wait to sign out for leave on a Friday. Well, the night before my flight, a snowstorm blew in and shut down all of the Puget Sound areas. My friend Kelley was able to drive me to the airport in his four-wheel-drive truck, and driving on an interstate when there was no visible road for miles on end was quite the experience.

Upon being dropped off at the airport, it was evident that no one was going anywhere for the foreseeable future, so I quickly phoned him to catch the ride back. I apologized for the nonsense and spent that holiday season alone in my shitty apartment because the airport didn't open back up for a while, given it was the most snow they had seen in a decade.

My birthday is in late December, so I couldn't enjoy the cake Celina had custom-made for me, with expectations that I'd arrive. Overall, it was an awful month, though I was building momentum for the future.

CHAPTER NINETEEN

EVERYTHING BECOMES CLEAR

After the new year, when everyone got back from leave, the entire brigade shipped out for Fort Irwin, California, to spend the next month at the NTC (National Training Center). NTC is where units getting ready to deploy get "certified" for deployment.

NTC was a good exercise that showed the lack of communication and structure that the unit had. All in all, it was a month-long joke. We spent a week in the cantonment area to prepare our vehicles and equipment for two weeks of the actual training exercise. When that ended, it was one final week back in the cantonment area to get our vehicles and equipment packed back up and put in shipping containers bound for Fort Lewis. We spent the first-week getting gear ready and being bored. I did paint my Barrett, though, which was probably the most significant task I did the entire week. It took maybe forty-five minutes to do.

After the week of nothing, we rolled into the two-week field problem; they told me that I wouldn't be used as a sniper because the tankers needed people to help them out on their missions.

It turned out that most of the missions done did not include any of the sniper team. The real reason I wasn't used or wasn't operating as part of the sniper team was that the entire company was pissed at me since I had just gotten back from SFAS.

One of the craziest and most absurd things happened to us was when they put me on baggage detail on the flight out of Washington to California for NTC. A sergeant from a different company came up and asked the three of us if we went to SFAS and then asked how it went. As if he cared! Then he proceeded to tell us that was why we were working on the baggage detail.

Though they didn't say it, it was apparent that they were trying to punish us for trying to better ourselves. The detail wasn't bad, though: we didn't do much, and the baggage detail also got to sit in the business class of the aircraft we flew on to California.

I didn't mind NTC overall. It was great because I got to see how messed up, lazy, and stupid our leadership was. The tent we slept in was big enough to fit our entire company without bunks. We slept on cots, which aren't too bad—better than the floor, although we also spent some nights sleeping there. The motor pool at the Forward Operating Base (FOB) was a good mile away from the tent, and most nights, we were so tired that we preferred to sleep in the vehicle than walk to the tent and waste precious sleep time. One morning, we were supposed to wake up at 6 a.m., but a sergeant that two other lower enlisted soldiers flipped out on decided to wake us up early that day because that was the last word he got. It turned out to be wrong, and Jack completely lost it, telling him that he was worthless and stupid. That sergeant then tried to explain why he did it (which would be wrong or bullshit either way).

"Shut the fuck up," Jack then screamed. "Just shut up!"

Jack's outburst was gratifying because when those two privates had flipped out on him, they knew the repercussions. Those two, though, because they were lower enlisted, caught shit and got in trouble. It was much easier for us to get reprimanded.

Everyone in the platoon and even the company knew which of us were competent and not, making it hard to understand why the command didn't take corrective action towards those of any rank.

There was one night, after an entire day of being out. I was looking forward to getting some sleep. That was when the sergeant who everyone flipped out on came up to me.

"Hey Perez," he called in a friendly tone. "Can you do me a favor?"

"What's up, Sergeant?" I asked hesitantly.

"Can you go to the tent to get my gloves and helmet? I'll buy you a gut truck[2] in the morning."

With my sleep deprivation and anger at the mismanagement of the entire company, I almost flipped out on him. But I took a deep breath, kept my cool, and calmly said no. We both knew that his request was bullshit, so he sulked and walked away.

It seemed NCOs liked to abuse their power now and then, some more than others. I've seen the same NCO ask his privates to do things that he is completely capable of doing but was just too lazy to do it. Their most mind-numbing move came when they ordered the sniper team to overwatch the town on the rooftop.

We sat on top of this roof on and off for three days, and on the third day, they decided that maybe it would be important for the snipers to get a High-Value Targets (HVT) list. This list is

---

[2] Gut Truck was the moniker for food trucks that sold us packaged food. Not like a typical food truck of today!

paperwork with pictures and names of "important" people to grab during missions should you come across someone listed. Things like that would be essential from the moment we stepped foot up there. After receiving the list, we looked through it and identified people we had seen before.

While on the roof, people (the role players) kept walking around the barrier to our combat operations base, and there was even a guy outside the fence taking pictures of us and our base. We tried to tell our "intelligence" guy (as in, anyone who would listen) about the people acting shady, but they told us not to worry about it. We didn't worry or bother to say anything to them after that since they wouldn't listen to us. At the end of the field problem, the cadre at NTC told us that the guy taking pictures of everything could be a possible sniper, and during the role-play, an enemy sniper was taking shots at us, but no one could identify his location. We found out where he was, got a team to move in on his position, and wound up getting him, but there was a catch. We didn't inform the cadre that we would move in on his position when we knew he was there. We acted on our own and got him, but everything was out of play since no cadre was around, so the sniper still got away.

One night, our company got wind of an insurgent wandering the streets. Being the high-speed soldiers we were, the guys hopped in their vehicles and surrounded the town while some of us walked in. Moving around the city, we were looking for someone, anyone. We ran into a cadre in what was mostly a ghost town, and he talked to Jack.

"You guys are chasing a ghost," he said. "Your time would be better spent getting rest for the next actual exercise."

We figured as much but still had to play the game. It went on for the next few hours, way longer than it should have. That is most of what I can recall from the time training at NTC, and during the breakdown week getting ready to go back home, there was one final incident. One of the mortarmen in our platoon lost a piece of his laser tag training gear we use. It's called MILES gear and uses infrared lasers on our weapons that make them laser tag guns when we shoot blanks. It's pretty much as practical as it sounds. He had lost it in the field and didn't tell anyone or notice until it was too late. Instead of taking responsibility for his shortcoming as a sergeant, he ordered his soldier to go over to another company's area and try to steal the item from one of them.

In the end, he got busted. The sergeant got into a lot of trouble and was no longer going to make any more rank in the military. He was moved and deployed with a different company, which brought the Mortar sergeant from headquarters over to our platoon to replace him. He seemed to have a better moral compass than the old mortar section leader, but I didn't get to work with or know him very well.

When we got back to Fort Lewis, the intention was to work out hard for the next few months leading up to deployment. We went to the gym daily, and one morning we were doing circuit training.

At the pull-up bars, Jack told me, "Perez, when you do pull-ups, don't use any energy coming down." He learned this technique in the Marine Corps, where they regularly do pull-ups. In the Army, or at least in the company, pull-ups weren't something we did too often.

That was one of the last times I worked out since I was gone again a day or two later for the final brigade-level sniper training I would do with Mr. E at Yakima Training Center (YTC) in Eastern Washington. I worked out hard because I was getting ready for war, and while we were there at YTC, we didn't work out at all, which I was thankful for as my entire body hurt and I was extremely sore from the intense workouts we did before leaving.

We got orders to deploy to Afghanistan instead of Iraq about a year prior. President Obama wanted more troops there. These orders changed a lot for the unit, like having the people who were formally learning Arabic to get ready for Iraq take a crash course on Pashto.[3] We allotted a lot of time using Rosetta Stone to try and learn Arabic for the rest of us at the company, so we spent all that time bullshitting our way through the courses for nothing.

---

[3] The Iranian language of the Pashtuns, also spoken in northern areas of Pakistan, that is an official language of Afghanistan.

CHAPTER TWENTY

A PART OF THE MACHINE

We were to be at YTC for a week, and as the pain and soreness left my body, I realized that my right shoulder still throbbed in pain. It hurt like hell. I knew as the days passed and as the rest of my body healed, there was something seriously wrong. I had such intense pain; I couldn't throw anything or even make a throwing motion without feeling discomfort. It sucked horribly, but I toughed it out while we were out training.

When doing pull-ups, I must have followed the advice of Jack a little too well. I used no energy and just my body weight to come down, which tore up my shoulder. There were some good times during that time training in Yakima, though.

My friend Clint was the sniper section leader for the battalion. His wife and mine drove up, and we went off post to enjoy some real food and watch a movie. It was a good time, and upon returning from training, I went straight into the bathroom to undress and took a good look at my body in the mirror.

I was shocked: my right shoulder was at least twice the size of my left, and it was throbbing with pain. It was only about four months until deployment, and I got X-rays not long after coming back from YTC. It hurt so bad that all I remember from being home those days was when I watched the NBA playoffs on my couch, pounding beers to try and kill the pain since they wouldn't give me anything for it. Every morning, I'd pull some bullshit somewhere, usually the barracks. Sometimes I'd run around making appointments that had to get done before deployment.

Finally, I received my X-rays. Afterward, I sat in the medical examination room, and the doctor came in, looking uninterested in anything as he flipped through my chart.

"I'll call you when I get the results," he said. "That should take around three or four days."

Well, two-and-a-half weeks passed, and I still got no call. At this point, I was too frustrated and curious to keep sitting around, so I went to sick call to see why I never got the call. It turns out, the worthless Army doctor went on leave and didn't tell anyone to call me. Another Army doctor had to see me, and it turned out I had a hairline fracture with a possible Hills Sachs deformity. I didn't know what that was, and apparently, no one else did, because the condition was never explained to me by a medical professional.

The doctor who looked at my X-rays didn't know about the injury, so they called a higher-ranking Army doctor to look at the X-rays and read the notes that came along with them.

"Well," he said after evaluating my notes. "I don't know what a Hills-Sachs deformity is, but don't do anything for six to eight weeks, and you should be fine."

I wanted to jump out of my chair and shout, "Are you fucking serious, dude?!" Instead, I grabbed my stuff and huffed out of there.

They put me on a two-week do-nothing profile, better known as a deadman profile. The dumbass who initially looked at my x-rays that day told me that I could still do sit-ups, even though she admitted that she didn't know what the injury was. Since then, I've never had long-term pain like this shoulder injury before.

Sometime during all this chaos, I was going through, I had Celina fly home to Texas because I had no idea what would happen, and I wanted us to leave on our terms, not because of the dumbass units. I had to fight to get her on that plane back home, and after she left, I started seeking out cannabis. I still wasn't getting any pain meds. Drinking wasn't my forte, so I wasn't going to smash alcohol on a nightly basis like I used to.

Cannabis was easy to get since I knew civilians, and I smoked that shit daily because I knew they weren't going to drug test us before deployment. I knew they wanted everybody they could get overseas, and drug testing everyone this close to deployment was sure to keep more than a few people from going. I smoked weed every day, and since everyone sat around for the four months before we deployed, I made a deal with Jack. He never stuck around the company during that time because he knew it was stupid to sit there doing nothing, so he went about his business, and I asked him if I could do the same.

"Hey Jack," I said, keeping my voice low. "Let me go fishing. It's right up the road, like five minutes away. If you need me back to the company, I can haul ass back over."

He agreed, and at some point during that time, I also asked him if I could go to our friend Chester's apartment, which was right off post, roughly ten minutes away. It was a privilege to have that apartment as a place to smoke so close to work; it's what kept me sane during those crazy times. I did the majority of my pot-smoking during the day there, and I probably spent more time there than I did fishing.

One day though, after coming back from breakfast, I walked up to the office area and went into our platoon office. My heart sank as I walked in and noticed the giant stack of small cardboard boxes, the same ones they used for the urine samples. I spent that entire day freaking the fuck out; I thought for sure that I fucked up because it was just a matter of time before they called to drug test everyone. As the day passed, my paranoia grew as I waited in anticipation for the call to line up, and I started drinking excessive amounts of water. As the day was nearing an end, I inquired as much as I could about what I saw without trying to make it evident that I was worried about it. Someone eventually informed me that Albie ordered someone to throw them away and that there would be no drug test.

I breathed a huge sigh of relief. Holy fuck, man, I was right! I knew this would be the case, and I smoked out hard that night because I knew I was good to go! I rode out my two-week profile, not working out but doing stuff for deployment (getting shots, making wills, etc.). Pissed and still in pain because of the lack of real meds, I went to sick call to bitch and requested an MRI and some actual pain killers.

This time, I got lucky. I saw an excellent Army doctor who put in my request to get an MRI, but he still wouldn't prescribe me any pain killers because, according to him, "Narcotics won't help your shoulder."

BULLSHIT, I thought. A narcotic pain killer wouldn't have helped my shoulder! I took the doctor's advice with a grain of salt, satisfied that I'd be getting an MRI done. I was confident that it would show that my shoulder was seriously injured, and then they'd have to do something to help.

Another two weeks passed, and finally, it was time to get the MRI done. Three to four days passed before I went back to the sick call to get my MRI read (which happened to be the day before deployment), and it turned out that I had SLAP tears and possible Hills-Sachs deformity. I was hoping the new army doctor would know more about Hills-Sachs deformity than the stupid, worthless Army doctor who read my X-rays last time.

I was wrong. She told me she would send an email to orthopedics to see if I would need surgery immediately or if ortho would let me deploy the following day.

"They should call before lunch, and if they don't, give me a call so I can take care of it," she said.

Before leaving post, I went to the company because I intended to leave my truck on post during the deployment, so I was hoping to find someone to explain my situation because I still needed my truck to make my MRI appointments. I walked into the bay doors, and I was surprised to run into Newt, my platoon sergeant immediately. I showed him the paper that said I needed to talk to ortho in case I needed surgery.

"Well, Albie said that we are taking a surgeon with us," Newt replied. "Albie said you can have surgery in Afghanistan." Once again, I was blown away by the ignorance of this place and just left before I exploded and said things that I would later regret.

I headed home and continued to pack, not knowing what would happen and waiting to receive the call from orthopedics. Lunch rolled around, and still no call, so I called the doc to inform her that I hadn't received a call.

"Go ahead and call Tricare to set up an appointment," she said. "You'll need a council to get an answer from ortho."

A lot of good setting up an appointment was going to do, considering that deployment was the following day.

Around noon I called Madigan, the hospital on Fort Lewis, and tried to connect to ortho. It rang and rang, and then I got the answering machine. I decided not to leave a message the first time because I planned to call again a few minutes later. I assumed that they were likely busy. About fifteen or thirty minutes passed before I called again. Still no answer, but this time I left a message about how urgent it was that I get an answer that day.

An hour passed with no call, so I called and called for two hours straight, and no one ever picked up. I even called the hospital service line to see if they could run someone by orthopedics to tell them to answer their phone, but they didn't cooperate and requested that I continue to call. The hospital secretary I spoke with could tell that I was peeved and asked if I would like to talk to customer service to leave a complaint about the ortho office. Usually, I don't care to waste my time doing anything of the sort, but in this case, I decided to. She connected me to the customer service, and it rang and rang and rang until I got the godforsaken answering machine. I practically threw the phone against the wall to hang it up. What a day I was having!

I had to wake up early on my day off, the day before deployment, to go and get my MRI results. Ortho never called me or answered my calls, and to top it all off, I got the answering machine for customer service. I was on edge, ready to lose my mind. I was in a shit mood, and it hurt me that Celina had to see me like that. She flew back a few days before deployment because she knew I was in a bad spot and wanted to be there for me until I was sure what was going on. These were some of our last days together, and the Army had me going insane. I was usually a fairly patient person, but I can only deal with so much on any given day, and it seemed like that day, the entire world was against me.

I wasn't sure if ortho closed at a particular hour, and since it was only around 3:45 p.m., I decided to go to the hospital to give them a piece of my mind. But there was a problem with that. I recently lost my ID and couldn't get on post. I called Tracey and asked him to meet me at the gate to get my wife and me on post. While he walked to meet us at the entrance, I continued to call ortho, and finally, an actual human being picked up.

Pissed off, I started to tell them how I'd been calling all day and how urgent it was for me to go in and talk to a physician. The man on the other line tore my world down. He told me that they got the email sent earlier from my care provider, but she needed to make an actual phone call.

"Ortho is not a walk-in clinic," he said. "The woman who did your MRI has to talk to a physician to plead your case to get seen."

My heart sank to my stomach, and Celina started crying because she didn't like to see these things happen to me. I met Tracey at the gate, gave him a ride to the convenience store, and apologized for making him walk. I proceeded to leave post and tried to give my "care provider" another call.

Just as I figured before making the call, she didn't answer. I left a voicemail, letting her know that she needed to call ortho herself to talk to a physician because they weren't a walk-in clinic. I never received a callback. I decided that once again, I'd wake up early and go to sick call the following day to explain in person the procedure needed to get seen by ortho.

Like every weekend, I didn't shave during that three-day weekend but figured that if I came in yesterday a little scruffy, another day of that wouldn't make much of a difference.

If I had gone in uniform, I would have shaved for apparent reasons, but I didn't think it was a big deal since I was going in civilian clothes. I was standing in line when a medic sergeant told me to get back. I backed off a privacy line and waited my turn again when the man behind the desk told me to come forward. Meanwhile, the sergeant who told me to back off the

line was talking to someone else but saw me walk forward and told me to get back again. It seemed like he woke up on the wrong side of the bed, so I waited, and finally, that medic came up and started to bitch to me about my scruff.

I admit my beard was kind of long, but seriously, I was there the day before, and he saw me but didn't say anything. It just goes to show that when a sergeant gets a wild hair up their ass, they take it out on anyone they can.

"Why didn't you shave this morning?" He screamed, just inches from my face.

"It was a three-day weekend, sergeant," I responded, my face blank. "I was going to shave before I showed up for work."

"It doesn't matter; you're a soldier 24/7."

He told me to leave and talk to the medics from work because apparently, they could help me. At the same time, he suggested that I shave before I go because his platoon sergeant would ream my ass. I made a quick detour to the barracks to cut, then headed over to HHC (Headquarters) to the senior medic's office. I talked to the medic platoon sergeant, and he called the captain, who was, I'm guessing, the only one who could access my MRI notes.

We waited for an hour, just bullshitting around, and while in the platoon sergeant's office, I saw a familiar name on the wall. It was the sergeant who bitched me out earlier about not shaving. The note was a reminder to weigh and tape him at the end of every week. It made me wonder. Did he seriously bitch at me for not shaving when he wasn't the perfect soldier himself? He was trying to act tough; that's one of the biggest problems with the army, hypocritical sergeants. Why ask your soldiers if you can't even perform the tasks yourself?

I sat there for an hour when Captain C finally arrived to talk to me about my shoulder. I proceeded to tell him the story from the previous day and said to him that I didn't know how the injury happened and that I was working out after NTC and was sore, but when my soreness went away, my shoulder still really hurt.

He pulled up my MRI notes and saw that it said I had a SLAP tear and Hills-Sachs deformity.

"Well," he said. "I don't know what a SLAP tear is."

That's when I sunk back in my chair, angry. Did Army doctors know anything besides obvious shit? Then he decided to go on

Google to search for what it was. The auto search's first suggestion was SLAP tear surgery.

This result was a relief because I figured that I would need surgery. Captain C called ortho, and they answered on the first try, but they told him that they had to go to the ER and call him back.

We went through a range of motions to see how my shoulder felt. It hurt, and I told him exactly where I felt the pain. After that, he tested my strength by making me put my arms out and making me resist his forcing them down.

"Well, you're strong," he said. "I think that you'll be able to deploy, but we will see what ortho says."

I wanted to tell him that I knew I was strong, but if your doctor isn't a complete bitch about his job, then you'd get adequate care of some kind. Instead, I told him the ridiculous story of Albie saying that we would take a surgeon with us to Afghanistan to see his opinion of it.

"Well, actually, they are talking about me," he replied. "I am a surgeon by title, but not an actual surgery surgeon."

Though I was skeptical of the entire situation, I remained silent. He said he would call me once ortho called him with the word on my surgery. I went home to see Celina already asleep. I sat next to her on the bed, and she woke right away. I told her I still had no idea what would happen to me and the two of us.

I rushed to finish packing and threw most of the trash out of my apartment, and while doing so, I missed a call from the captain but nervously listened to the voicemail. I don't remember what it said exactly. What I got from the message was that I did need surgery, but postponing it was okay since it was deployment day. I rushed to pack, took stuff to my storage unit, and got things together. 10:30 a.m. rolled around, and I had to head to base, still unsure if I was deploying because they just told me to stage my bags at the company.

I had to have a friend meet me outside the gate because I still didn't have an ID. On base, I changed from civilian clothes to my uniform in my truck and walked up to the company. I saw Jack walking up to me.

"You lost your ID?" he cried. "When did you do this?"

"I told you last week," I said. "I lost it either at Albertsons or at my storage unit."

"Did you send it to me in a text or did you tell me in person?" he said in an angry tone.

"I told you in person, Jack. Right, Chester?"

Jack snapped right back after turning towards Chester for a millisecond for confirmation.

"Everyone thinks you're trying to get out of this deployment!".

He was likely mad because he couldn't get me in trouble. Not only did I tell him in person last week that I had lost it, but I also had Chester as a witness. He lost his composure when Chester confirmed that I had told him because, at that point, I was only partially to blame.

"Now Newt wants you because he's gonna take you to get your new ID."

Being told that everyone thought that I was trying to get out of deployment was the biggest insult that was ever said to me, especially after I helped build that place from the ground up. I accidentally lost my ID, told my team leader about it the week prior, and because I got my MRI results the day before deployment and was told that I need surgery, I'm trying to get out of the deployment?

They acted like I intentionally did all this, and the thing that bothered me the most was that they thought I was trying to get out of deployment. I had been waiting to use my sniper training for years; if I wanted out of deployment, I could've gone AWOL like a bunch of other soldiers did.

CHAPTER TWENTY-ONE

6 HOURS NOTICE

Celina and I ran around post to get my new ID for the next few hours. Looking back on it, I would have much rather stood there with her, just hugging, kissing, and talking.

Since I was in so much pain, I intended to bring weed on the trip to Afghanistan to smoke along the way, but I ended up burning through everything I bought before leaving and couldn't get more. I was disheartened but figured that I would get some once we reached our destination.

We finally loaded the buses and were off to catch our flight overseas. We bullshitted around waiting in boredom, but eventually, we got on our plane headed for a world of hurt. We made two stops in Maine and Germany and eventually got to Manas, an Air Force base in Kyrgyzstan. It was a beautiful place—not very hot during the day and not too dusty. The chow hall stayed open twenty-four hours, and they had decent places to use the internet and phones.

We stayed at this location for four days, and the first day, everything was great. The next day, not so much. We piled our weapons on a bed on that day because we weren't allowed to walk around with them. The leadership made a guard roster for the lower enlisted personnel to watch the weapons, an hour for each person, usually the amount of time for a soldier on guard shift. It turned out that one of the soldiers on guard turned his back to the weapons for a minute, and Albie caught him. Needless to say, he bitched out the sergeants.

"They're going to pull guard on the weapons!" he cried. "E-4 and below don't know how to do it."

In the Army, "Shit rolls downhill," so when someone gets in trouble and subsequently gets their supervisor in trouble, the "supervisor" usually decides to "smoke" or "correctively train" the soldier who messed up. Strangely enough, that didn't happen, but the sergeants from the platoon got us all together for a huge lecture.

"All of you," they said. "All of you fucked up, but it's not that big of a deal because it wasn't the middle of the night, and there were other people from the platoon around."

No one blamed the guy on duty because pretty much all of us had our backs turned towards the weapons while on guard, and I couldn't believe that there were reasonable sergeants among us. But now that sergeants had to pull guard on the weapons, they wanted everyone to check in every hour. That was dumb. This new practice gave us minimal time to do anything without walking back and checking in for absolutely no reason. The internet was slow, so I might have been able to check my email and send a few messages to Celina in an hour. At that point, I would go back to check-in.

Another inconvenience about the base was that no bags were allowed in many places, so I had to carry my laptop and anything else I wanted to take. On the third day, I used the internet and watched a movie at the recreation center. The hour was coming up, and Tracey was about to go and sign in, so I asked him if he could ask the sergeant on guard if it would be okay if I stayed where I was, and if it weren't, then I'd go to sign in. I waited, and Tracey left, the hour passed, and he never returned, so I figured everything was okay. I opted to continue using the internet without care or worry in the world. I was enjoying myself.

Suddenly Timo came in. "You need to go to the tent," he said. "They are pissed at you."

Oh god, I knew it would be bad, but I had no idea of the storm that was brewing. I left my laptop there with a friend and told him that I would be right back. I figured if I told them my intention, they would understand, and everything would be okay. Of course not, since this was the 5th Brigade! Timo and I walked back together.

"So they are pissed, huh?" I asked quietly.

"Yeah," he said. "Jack is pissed at you."

"Really?"

It was surprising because he was usually patient, but he was probably waiting for me to slip up with all the bullshit that happened in the last week. I showed up and approached the sergeant on guard to tell him what had happened.

"Nobody told me anything," the sergeant said, "but Jack wants to talk to you."

Talking wasn't really what he wanted, and a few seconds later, he walked through the door near the guard spot and immediately told me that I had sixty seconds to run to the computer and internet spot to grab four bottles of water. I knew this would suck because I hadn't worked out in months and technically wasn't supposed to because of my profile. But I ran to grab the waters and was back in way under sixty seconds.

Then, the worst began.

"You have two minutes to go and put on your ghillie suit," he told me. I ran to my bunk but realized that I left it in a bag in the caged area where they kept the bags we didn't need.

"Jack," I said when I ran back. "I don't have it; it's packed in a different bag."

"Put the waters in your pockets and run to the left side of the smoking area."

I ran my ass off and waited there. Finally, Jack arrived.

"Get in the prone," he said. I lay on the hot sand, belly down. "Start low-crawling. And I don't want to see any dust!"

He wanted me to move, slow as is the nature of a sniper, so I crawled. While I did this, he lectured me about how we are expected to be on time as part of a sniper team and many other things. The gist of what was said was that we are expected to be model soldiers, which is understandable because there are only three snipers in a company of 150-200 soldiers. I tried to explain what happened, but there are no two sides to a story when a sergeant goes over the edge. It is what it is, so I had to crawl for about fifty yards.

After reaching the concrete, I had to sprint to a wall and back, which was seventy yards one way in less than forty-five seconds. I made it, then had to sprint again in less than thirty seconds.

I'm not sure if I made it the second time because the first sprint with boots took a lot of energy out of me. I had minimal stamina since my injuries prevented me from working out for a long time.

Again, I had an additional forty-five seconds to sprint back to the wall, and at this point, I started dry heaving to the point of puking. When I got back to Jack, I was instructed to crawl to the concrete again. While crawling, I puked and puked and continued to crawl through it. Crawling through my bile was not an issue for me. I had no problem with it after rolling through puke during SFAS, both mine and other soldiers'.

At the end of it all, I was tired, and it wasn't a good day to decide to eat the cheeseburger meal for lunch. I was soaked with water, sweat, and puke. I'm just glad it happened prior to me doing laundry. Jack was glad that I didn't quit, and he asked me during the "corrective training" session if I wanted to.

"No," was the only word I said in response. I knew deep down that I wouldn't quit, ever. I would pass out, or the session would stop. That's what these people couldn't understand about me.

The human body can only go so far until muscle failure. Once your muscles fail, they can bitch and cry for someone to do more, but it's not going to happen. Not in a "pretty" fashion anyway. It turns out I scraped my knee pretty good while low-crawling over the sand and rocks.

Great, I thought as I saw the blood oozing from my knee. *An open wound on the right side of my knee and we will soon be in a third-world country.*

We left for Afghanistan the following day, but I learned something vital during my time there. Shit was rough, and I was pissed that I couldn't score any weed before I left. Then, as if the cannabis gods could hear me yearning, one of my close brothers came to me with the green goodness. I couldn't believe it! I'm sure I brought it up and mentioned that I was pissed that I smoked my stash and didn't bring any with me because it was prime smoking time. He didn't have anyone to smoke with and knew he could trust me, so he pulled out his cigarette one-hitter and loaded it up.

We got nice and toasty that night and exchanged weed stories. I told him I'd only been smoking for the last few months, shortly after my injury. That's when he told me that he had been smoking the entire time in the garrison because he was the designated urine test administrator, and they don't get tested. How ironic that the people tasked to give the drug tests are the only ones immune from taking them! I was blown away at this fact and was jealous and happy for him because he had a better outlet, unlike the alcohol most of us were using to drown our sorrows in.

We found more smoking buddies during this time, and I leeched off him the entire trip. I'm forever grateful because that weed helped with my shoulder pain and the mental torment I was enduring.

CHAPTER TWENTY-TWO

HANDICAP BY DISARMARMENT

On the Air Force flight to Afghanistan, all I could think about was how the Army screwed over every single person on this plane, in one way or another. Whether it was multiple times or just once, The U.S. Army has done wrong to them all, and for me, that means this system was incredibly broken. I'm sure that everyone in each branch of the military could write a book about how the military has screwed them over or how they loved their time in service, and it was the best choice that they could have ever made. There were a few things the Army did well for me at the time. They gave me the G.I. Bill, taught me to appreciate life, and opened my eyes to how broken our systems are.

I realize that I tend to provide others with a bitchfest—anyone is welcome to criticize me or praise me, but I don't care. I just want the public to know that the people who ran that show were beyond clueless. They never knew what was going on, and it was deplorable because many lives were at stake.

All I hoped for was to do the tour in Afghanistan and return with a years' worth of material to piece together and publish my works. As I experienced it, Afghanistan was hell on earth for American soldiers and soldiers from all over the world. Service members were dying almost daily there, and I knew that some of us on this plane wouldn't be coming back. It's sad because not only did this unit and the Army make their lives hell, the soldiers and their families made the ultimate sacrifice by giving their lives. And what is the cost? What price do you put on a life? The majority of the people here make less than $3,000 a month and are very unhappy with their quality of life.

The only thing that kept us sane was knowing that we were all hating life and going into hell together as a unit. On July 21st, we finally got to Afghanistan, and it was almost what I came to expect. It was so hot and dry out there, and to exemplify the horrible atmosphere, the best experience was driving by the "Shit Pond." It's an open septic tank—literally a shit pond—and contains the most rank smell that has EVER hit my nasal cavity. Not only does it stink, but it's open, giving flies a nice place to take refuge.

The whole base didn't seem very well planned out. Something that got under my skin was that the people running the show had pallets of water sitting directly in the sun. With all the money that the government spends on war, you would think they would at least throw a tarp or something to provide some shade for the water that we drink. You could seriously make coffee with the water straight from the bottle. Even at night, the water was still boiling from the direct sunlight and 100-plus degree Afghan weather.

One of the first days we were there, I was just off a two-hour detail in the 100-plus degree heat. I entered shipping containers to get serial numbers for weapons and equipment when I heard about another detail for tomorrow at 7am.

That's when I thought, *Thank god I'm doing this detail now; that way, I won't have to go if we get to sleep in.*

Once we were done and returned to our somewhat cool tent, the person in charge came up to me.

"I think Albie has it out for you," he whispered to me in passing. "You're on the 7 a.m. detail."

I figured from that point, I would probably be on every detail he could've put me on during the deployment. Sure enough, it wasn't long before he confirmed it out loud. "Every detail will be however many guys are needed and Perez." Albie was like a junior high kid with too much authority. He still had a grudge against me for going to SFAS.

On July 22nd, 2009, we went to the motor pool because the vehicle crews had to get equipment from the shipping containers for their rides. They somehow dared to make us, the snipers, go down to the motor pool. None of us had ever messed with a Stryker[4], and we didn't know where shit went or how they wanted things.

When we got there, we just sat around and did nothing, and thankfully the recently invented smartphone helped keep me entertained. We only wasted six hours of our day, and there were many things I could have been doing. Instead, they opted to make us go down there and sit on our asses. We could only help a little when they needed to move stuff, but it was a waste overall. And as usual, we were the last platoon to leave the motor pool and had to pick up everyone else's mess.

Before we could walk the mile back to our tent, we got together as a platoon for our Platoon Sergeant, Newt, to put information out to us.

---

[4] The Stryker is a family of eight-wheeled armored fighting vehicles produced by General Dynamics Land Systems (GDLS) for the US Army. The Stryker was conceived as a family of vehicles forming the backbone of a new medium-weight brigade combat team (BCT) that was to strike a balance between heavy armor and infantry.

"I hear that we are the platoon that complains the most," he told us. "It's hot. I don't wanna be here. I hate this place." He turned to us for encouragement as he continued the mockery. "We are always the last ones off. We always have to pick up after the other platoons. I'm tired of getting bitched at because you all complain!"

The ironic thing about it all was right before this lecture, one of the squad leaders complained about being the last ones off and having to pick up after the other platoons. The thing he said next, though, completely blew my mind considering the area and situation.

"If something is bothering you, keep your mouth shut, and don't say anything!"

Internally, I was scoffing. Was he fucking serious? We were in a combat zone where people lose their sanity all the time from the shit they see and do, and this guy is telling his soldiers to keep their mouths shut if something is bothering them.

The way I interpreted his advice was, "Hey if you're depressed about being here, don't say anything and let it build up inside. That way, it can eat at your soul until you're so depressed you off yourself."

It was a good thing I'd built up enough mental strength, so the shit they did and said didn't bother me too much. In fact, it wasn't a tremendous bother for me to be there in the first place. I came to terms with the fact that I was probably not going to go home alive, which is why I tried to board the SFAS life raft to get off the sinking ship. But I acknowledge that I'm lucky in my ability to be unbothered, and not everyone is simply given the ability to deal with shit like this.

Even during NTC[5], people were highly pissy—sergeants were constantly at each other's throats. It was a year-long deployment, and we hadn't yet finished the first month. The leadership had their heads shoved so far up their asses that they weren't going to see the signs of their soldiers' depression. The Army expects the peers of the depressed soldiers to recognize the signs, and that's fine, but as a leader, I would talk to my soldiers every night to make sure they were doing okay.

Something interesting also happened around this time—they took away my sniper rifle. Why would they take away a sniper rifle from a sniper team member to give it to someone less skilled and proficient in the weapon? Well, it might make more sense if I told you this was Albie's call. His reasoning? That the sniper team had too much firepower. I've been training for over two years for this, and they took away the most important tool I had. They might as well have just kicked me off the sniper team.

The Taliban were an intelligent enemy; this became more apparent after attending the counter IED class. The Taliban would dig through the trash we throw out and then use our equipment and waste against us. I was amazed that people could do such things with minimal education.

It saddened me to know that my friends may not be going home, and I told myself that I would do everything in my power to ensure that nothing happened to anyone. I was only one physically and mentally broken lower enlisted man, though, so I was limited in what I was able to do.

---

[5] Fort Irwin National Training Center (NTC) is a major training area for the United States military in the Mojave Desert in northern San Bernardino County, California.

*****

The 3rd and 4th of August were interesting days. It started with Albie's typical shitty leadership—an excellent way to start any day. He told other platoons' soldiers to wake the snipers up because he decided we would go to the counter IED[6] lane that morning. It was good training since they taught us how to identify an IED threat with small indicators that the Taliban left. They mainly taught us that we would likely get blown up by all of their homemade weapons and bombs.

The classes went on throughout the day, and it was miserable outside: windy, dusty, and hot as balls. It also sucked that all the water we had to drink was one-hundred degrees. Noon rolled around, and it was time to leave the class and head back to get ready to go to a range that was supposed to happen at 1300. We got back and dismounted before the vehicle left to gas up. As soon as we dismounted, we sought shade from the sun that beats this place down daily.

We laid around for a few minutes under the front of a Stryker, then decided to head back to the other platoons' vehicles as they should've been there to go to the 1300 range. We showed up and recieved news that the range was now at 1500.

---

[6] IED's are improvised explosive devices that are bombs made of everyday materials. The lane was meant to help us in identifying these devices.

Trying to enjoy shade anyway we could

About ready to rip open an MRE,[7] I dropped everything except my weapon, and Timo and I walked to the chow tent to do the usual. We waited in line, washed our hands, went through the food line, ate in a typical hot tent, and then left. We hadn't had a decent meal all day and had been exposed to the elements and hot water during the class. It was refreshing to sit at the table, eat, and drink something cool.

The best way to cool our drinks was to leave them on the air conditioning vent in our tent. There were four of them in the tent, and they did a decent job keeping the tent bearable during the day and cold

---

[7] Meals ready to eat (MRE) are what soldiers eat in the field that are high in calories to fuel us in the field.

at night. The first day we got situated in our tent, I left some drinks on the vent and got told not to do it because "Albie said so."

It was just a matter of time before everyone started doing it, and the cool carbonated drink made me happy. Before going and eating chow, I was told to bring my primary weapon to the range, the .50 caliber Barrett sniper rifle. I had been eager to shoot it and see if traveling to the other side of the world had affected its zero.

At the Stryker, we were prepping to head out around 1315, a little late, which usually isn't the way things are supposed to work. I walked, without gear, to the motor pool, and it wore into me a little bit. I imagined the hell I would have to bring it back to the tent with all my gear on. It wasn't until we mounted up to wait for the company commanders' word when we got word that the range was pushed back again.

Meanwhile, we are all standing and sitting around in the blazing heat, waiting for this thing to kick off. We ended up having two soldiers go down as heat casualties, one of whom was my buddy, Tracey. They made me the squad leader of the vehicle I was on, while the actual squad leader took one of the heat casualties to the medics.

We continued to wait, and finally, we headed out towards the range, which was about three miles from KAF outside the wire. On the way there, we got lost. Three miles outside the wire, and we got lost. We eventually found our way, and when we got there, it was almost dark. We sat and waited for about fifteen minutes and were told to ceasefire to let our mortars shoot their mortar weapon systems. The vehicles fired their weapons to ensure they worked while we waited in them.

I found out later the next night that we were shooting at a minefield cleared for twenty-five meters. For the mortars to do their job, they had to run a metal stake out one hundred meters. They call it an aiming stake.

Axel was there with the company commander, and I talked to him about running out there. I don't know if they knew it was a mine-field, but they took the risk anyway and still sent him out, all because the CO[8] said that he would be okay.

According to Axel, as he ran to stick the metal rod into the mine-infested ground, rounds fired, kicking up the dirt. They sent him into a live-fire minefield to stick a stake into the ground to shoot a mortar, and someone was still firing.

After that little shindig, we were told to shoot three rounds to make sure our rifles worked, so it took us around six hours and three heat casualties to get everything done. We almost had one soldier killed, all to shoot about three rounds per person and maybe twenty rounds per vehicle. And this was only the beginning of what was to be a long year.

---

[8] Commanding Officer.

CHAPTER TWENTY-THREE

THE REALITY CHECK!

When we arrived at forward operating base (FOB) Frontenac, it was already night. I was standing outside our tent area with a few guys from my platoon. My idea of a joke was threatening their lives with a water pistol that I made all black back on KAF, and everyone was playing along, feigning fear with their hands up. That's when our Company Commander walked by.

"Hey!" he yelled. "What's going on here, men?"

"SIR!" I cried, standing up straight. "I'm just joking with them; it's a water gun."

"Hah," he said. "Can I see it?"

"Sure sir, no problem."

Though I handed the toy over, I thought it was strange that he still wanted to take a look. He wanted to confirm that it was filled with water and not lead. That was one of the last moments of free time we had before we began our missions.

August 18, 2009, was an interesting day. We woke at 4:45 a.m. because we were expected to be getting ready for missions at 7. I spent most of the prior night with the Charlie Company snipers at their CHU.[9] I enjoyed hanging out with them because we all got along well and joked around a lot. They, by far, were the closest-knit team in the battalion.

---

[9] Combat housing unit.

I left their CHU around 11:30 p.m. and headed back to my tent. There, we spent the afternoon rearranging since a platoon got CHUs and another platoon was sent to a different tent. I was glad because I finally got a bottom bunk near a power outlet. I set up my bunk after they were moved, and after two hours of slowly getting everything situated how I liked, I went to eat around 6 p.m.

When I got back from the chow house, I got word that we would be rolling out the following day and staying out for three days for our mission. Well, ain't that some shit. There is probably around a twenty-degree temperature difference between a top bunk near the edge of the tent (where I slept prior) and a bottom bunk. I got back from hanging out with the C. Co. team, took a shower, and began packing for the three-day mission.

All that's needed for a three-day mission is one uniform, five or six pairs of socks and shirts, and the typical hygiene and mission essentials. I knew I wouldn't get much sleep that night since we had a different mission that was supposed to be at 2 a.m. instead of 5 a.m., so I opted to have a caffeine drink with dinner instead of a caffeine-free one. I don't drink much caffeine, so I'm borderline intolerant. A single soda too late at night will keep me from falling asleep.

I managed to fall asleep around 2:30 a.m., and waking up at 4:45 wasn't too bad. I wasn't tired at all. It was weird, though, because that day from 5-6 p.m. was the fastest hour of my life. During that time, I didn't get to do shit. For example, when we were woken up at 8 a.m. for a 10 a.m. mission, from 8 to 9, I'd drag ass getting out of bed from being so tired. I was still able to brush my teeth, wash my face at the shower tent, take my laundry to get cleaned, pack my assault pack, and eat a can of oysters.

Our mission for the day was to drive to different areas near villages and climb up tall hills near their peaks to overwatch said villages and roads leading to them. It was strange that the hour had passed in the blink of an eye. Before I knew it, we were on the Stryker with 2nd platoon.

Before lunch, the Afghan National Army (or ANA) got intel of around fifty Taliban in a nearby village. Supposedly, an ANA officer visited the village, saw people packing, and asked what was happening. The Taliban told the villagers to leave, so of course, they did. Unfortunately for Timo, he felt sick the whole time and even puked in the Stryker.

After we got news of this, the ANA decided to take a lunch break. As we sat in our Strykers near the ANA compound, I decided to go to the mortar vehicle and talk with Axel. We sat around, talking about how there was probably no Taliban at the village and that it would be a yearlong NTC. We also talked about how worthless the unit was since they didn't properly employ the mortars and snipers. Even Jack was losing his patience!

This mission's icing on the cake was the last (and probably highest) hill we climbed near the ANA compound. It was a bitch getting up that mountain. We finally got to the top, and about ten minutes later, we were told to come back down. That was the fourth climb of the day and all before lunch, so Jack decided to change from our ACUs[10] and IOTVs[11] to Multicam and plate carriers, which are one-hundred times more comfortable and allowed us more maneuverability. They offer a little less protection, but I'll

---

[10] ACU Army Combat Uniform was the Army's everyday uniform

[11] IOTV Improved Outer Tactical Vest was bulkier body armor with side plates to protect under armpit area

choose speed over more weight any day. Jack was also the one who bought the gear for all of the team thankfully!

In Multicam, we rolled up to the village, ANA leading the way. We heard that ANA was in a firefight with the Taliban over the radio on our way there. An actual gunfight? Usually, the bitch ass Taliban just buried IEDs to blow us up.

Sure enough, we arrived, and shit was popping off. We pulled to the right of the hill where ANA was posted up, and then O'Brien asked Jack if he wanted to dismount the sniper team. Of course, he said yes, which I was all for.

Off into the firefight, we went. We ran out of the small hatch on the back ramp of the Stryker. I had faith that with my better equipment, training, and support that I'd leave this place alive, though I did know all it could take would be a lucky shot to kill me, and I had already accepted the fact that I might not go home from that place.

ANA fired out into the village and orchard. We continued towards the hill. As we approached the base of the hill, one of the ANA personnel was using a massive rock as cover while shooting his AK47[12] from behind it. I was watching him provide cover fire while we were running toward the hill, and after a few shots, he ended up shooting the rock that was right in front of him. I couldn't believe it! He stopped shooting because a piece of the rock he just fragmented caught him on the side of the face, and he was bleeding, so he called it a day. When I witnessed that, my concern switched from being killed by the Taliban to being shot by one of those idiots.

---

[12] The AK-47, officially known as the Avtomat Kalashnikova, is a gas-operated assault rifle that is chambered for the 7.62×39mm cartridge.

After pushing our way up the hill, we reached our first position. The entire time, rounds from the Taliban zipped overhead and ricocheted off rocks while we used the terrain as cover. Timo and I were using our M4s while Jack had the SASS. We were both catching our breath and hydrating, as I was no longer in shape and he was sick as a dog. He didn't have his water, and when we got to the top, I began drinking from my Camelbak when he asked me for a drink. When I was done, I tossed him the mouthpiece and began scanning the area for targets to engage.

We were there, staying low. Each man assessed the situation. That was when I noticed an ANA machine gunner about twenty feet in front of me, sitting in a mortar crater with the gun just barely sticking out of the indention. I admired the position, thinking that if I were still a machine gunner, that would be ideal as he was well covered in the low spot.

While admiring the position he was in, BOOM! A mortar round went off in front of us right by him. Holy shit, that was seriously intense! As soon as the mortar round hit, my head instinctively snapped away from the blast. The dust finally settled; I looked toward where the round hit and saw a bunch of shaken ANA men walking around in a daze, as well as an unmanned machine gun pointing straight into the air. It was manned before that round hit anyway, and the ANA personnel manning it was carried off by two of their men.

I was a slight breeze away from losing a limb and, more likely, my life on my first actual gunfight. That would've been some shit. I guess my bad luck wasn't that bad on that day, and my first thought after the mortar hit was *we should move because if the Taliban saw where it impacted, they would make it rain.*

I looked over to Jack.

"Hey," I said. "I think we should probably relocate."

In my mind, I envisioned the fighter who launched that mortar that almost hit us was about to fire for effect and take out that entire hillside. Thankfully, Jack agreed.

We moved towards the village, still on top of the hill, while under constant fire. The ANA was something else—they are the craziest sons of bitches I'd ever seen. They walked around without helmets or cover, shot wildly, and bunched up instead of getting in line. I wasn't sure if they were brave or just high out of their minds as their "tobacco dip" is just a mix of opium and hash.

A few of them were taking cover, but we had a theory on why they carelessly walked about while under fire. The Taliban didn't want to kill them, and I'm sure some corrupt ANA personnel gave the Taliban intel.

The ANA commander was there, leading us to the side of the hill visible from the village and orchard area. The ANA commander led us to a dangerous area and quickly left, which made Jack and I suspicious. The three of us were under constant fire while the ANA stood around, and they eventually decided to break contact shortly after we showed up.

We got back up and left the exposed area. We fell back to another rock for cover and took a lot of fire. Being the only three Americans on this hill with Multicam seemed to draw extra fire since only special operations guys wore this uniform at that time. We were pretty much begging to be shot at any time we wore that gear.

Wearing Multicam in the deployment days of ACU

As the fight went on, we began taking fire from eight o'clock to four o'clock from all directions. A Stryker rolled up to pick us up, so we told them to drop the ramp to let us in. The rate of the Taliban's fire seemed to have tripled when they saw us behind the vehicle, obviously trying to get in. Every Stryker ramp has a troop door that opens like a regular door, except its 300-plus pounds and about two feet by four feet at most. It was like straight out of a bad comedy watching that tiny door open instead of the complete hatch dropping.

I jumped in first, with some ease. Timo was right behind me: he was about five-foot-four, and I knew he would have trouble getting in, so I turned to offer my hand. Then, Jack pushed him in as I made my effort, so he flew into me.

The Stryker was packed—thirteen people in the vehicle with full gear. Strykers have the boomerang system, which tells us the distance and direction of rounds being shot at the Stryker. As soon as we got in, a computerized female voice called out.

"Shot three o'clock, shot eight o'clock, shot, shot, shot."

They were shooting from just about every direction, it seemed. We waited for quite a while in the Stryker, and while we were on the hill fighting, I saw our mortars emplacing their mortar system. It turns out that they never got to send up any rounds because the fire officer who had to clear them was scared to make a decision.

After all that went down, not much happened, though it was a multi-hour ordeal. It turns out we drove through a minefield that night when we went to link up with the mortars because they had our assault packs on their vehicle. The funny thing is, we didn't know it was a minefield at the time, and it was the mortars who informed us. We went to a different spot after we linked up with the company commander and got in his vehicle since there was more room. We were out all night until about 1 a.m., and it took forever because a vehicle broke down, so we waited for another Stryker to recover it, which took all night for some strange reason. After getting our bags on the Stryker, we finally rolled back to the FOB.

Meanwhile, I acted as air guard for two-and-a-half hours while Jack and sick Timo caught some Zs. While on the way, I was entirely out of the Stryker, laying with my torso in an upright position on top while air guarding. Like my comrades, I was tired from being up at 4:45am and the day's events.

The standard operating procedure for air guards, one I didn't follow, was to tie ourselves to the vehicle with a rope. Their reasoning behind this was to keep us from being flung from the vehicle in the event of hitting an explosive device, and I thought I had better chances of survival if I were launched. At some point, the commander caught me since I was entirely outside the air guard, hatch laying on top of the vehicle with my back propped up against the shit we had strapped down.

I went back to standing in the hole, but there was no way in hell that I was going to tie myself to the vehicle. I was hallucinating from being so tired. Being on top of the commander's vehicle was nice because when we arrived at the FOB (which felt like it took forever), he got to drive up to the TOC, roughly twenty feet away from the tent. We were told to sleep in when we got there, which was nice.

Upon entering the tent, I was relieved to be back to a place to relax. I walked to my bed, and it was completely covered with stuff from the guy who slept next to me. I walked up, grabbed the side of the mattress, and dumped the bed to get the stuff off. It was irritating to show up and have all that shit on my bed—I was so tired and frustrated. I sat on a tuff box to take a load off.

As I started taking off my boots, one of the mortarmen stopped by my bed and asked about my day. I gave a quick synopsis of the firefight and the day; then he began to tell me that 3rd platoon, who had been attached to B-Co the day prior, had lost a soldier.

"Confirmed KIA,[13] Bidzii."

I couldn't believe it coming from him, but he did say 3rd platoon's platoon sergeant broke the news, so I wasn't sure what to think. I went to bed that night uneasy and unsure of what happened. The guys from 3rd platoon were still out, and It wasn't until the following day that

---

[13] Killed In Action (KIA) is the term used for military personnel who are killed during military operations.

I knew for sure that he was gone. When I woke up, I looked over to 3rd platoon's area, and they were huddled up saying a prayer, and their platoon sergeant was telling them that they needed to continue on, that there was still a job to do. Their faces and body language said it all: Bidzii was gone. To think that he was never going to come back and that there was a good chance we ourselves maybe next on any given day. After the prayer, one of 3rd platoon's team leaders, we'll call him DB, came over to me and had a look of sheer terror in his face. He demanded that I give him my grappling hook since I was the only one with the foresight to bring something so useful. He said they wanted to use it and wouldn't need it since I was a sniper. That was always their fucking excuse.

I knew that he knew he was full of shit, and I was hesitant at first, but thinking about my brothers and the one I just lost, I handed it over to him.

I broke down walking out of that tent after honoring Bidzii and the other soldiers' sacrifice.

CHAPTER TWENTY-FOUR

CAN WE CHECK AGAIN?

August 22nd, 2009 started with being woken up at 8:30 a.m. by sergeant Rosales telling me that we had a mission at 9. A whole thirty minutes to get ready for the task of protecting Re-Trans, whose purpose is to drive to the top of a mountain and set up antennas to help us communicate better.

I got my gear with no time to do anything and stopped by Jack's bed to see what was going on. A few days prior, I turned in my laundry and forgot to take my dog tags out of my pants. When I got my laundry back, the dog tags were gone. The Afghani person who threw my laundry in decided to take off my dog tags and leave them in the restroom. Well, because I have the best luck, Albie was the one who stumbled across them. Albie gave them to Jack and told him to "smoke" me for losing them. Jack just told me to keep account of them, and if he asked, I got smoked.

I knew it was just a matter of time before Albie kicked me off the sniper team. At the time, I'd been telling our platoon medic that I was still in a lot of pain from my shoulder injury, but as long as I was on the sniper team, I wasn't going to complain about it. The moment Albie pulled a bitch move, I knew I could do the same right back at him. I worked too hard and been in that section too long just to be kicked out by a worthless First Sergeant who had a god complex. Jack came to me excited, and I wasn't sure what was going on,

"Perez," he said. "You'll never guess what the commander just told me!"

"What's up?" I answered. "Hope it's good news."

"Our commander said that we can do whatever the fuck we want! He told me he didn't know how to use us, so we're free to do as we want for missions."

"Damn! That's great news; we don't have to do what Albie tells us now!"

It was the best news I heard all year; I was eager to see how things would play out but wasn't expecting things to get too much better in my personal hell. I was also glad that our commander wasn't disillusioned by his position and made the right call because most commanders don't know how to employ snipers. The Army developed a course for officers held at Sniper School in Georgia to teach them how to utilize our force-multiplying capabilities.

On August 25th, 2009, two things were going on. One was a humanitarian mission to a local village to drop off medical supplies and offer assistance, so our commander brought senior battalion medics with him.

The other mission that we ended up helping with was just the usual: climbing mountains and patrolling. Jack pleaded with Albie to let us go with the commander since we were his asset. He could've brought up what the CO previously said, and he may have. I don't know if he did or didn't; I know I would have pulled that card as often as possible. However, Albie made us go with our platoon to help them patrol.

We all left at the same time to escort the commander's vehicle, and when we were at the staging point before leaving the FOB, I was air guarding in the vehicle in front of his.

Before leaving the FOB, I put on my "Everything Sucks in Afghanistan" morale patch. The morale patch was a round white patch sold on KAF that had different sayings on them to boost your mood or start conversations. The variations of them had a modicum of ways in which you could suffer in that place: diarrhea, sandstorms, heat, and malaria were some of the few options you could choose from. These things didn't bother me the way the fuck heads in charge did, so I used a black sharpie to cross out "heat" and cover the word with "COC" for the chain of command.

The driver of the commander's vehicle, Luis, was the same rank as I and a good-hearted guy. In garrison, he looked up to us in the sniper spots, and he would occasionally shoot the shit with us. At one point, they sent him to do sniper training somewhere; I think it was when they sent a group of guys to a school hosted by the National Guard. Luis got my attention and gestured to his shoulder, trying to inquire about what I had on since it wasn't "uniform."

I tried my best to gesture what it was, he just shook his head in confusion, and I waved him off to insinuate that I would show him when we got back from the day's mission. After a few minutes of waiting around, we headed out to the ANA police compound, which wasn't far from where the commander needed to go, and when we got to the compound, we sat for a bit while the higher-ups had a short meeting.

We left the compound as a platoon except for the mortars; they went in the commander's direction for a different mission. We made two visits to small villages in the area and headed to a third. When we got eyes on the village, we stopped for about ten minutes to use our optics to look before heading in and started working towards it after the reconnaissance.

A few minutes later, the radio broke. It screamed. "CONTACT!"

We didn't know who got hit, but it didn't matter. We went off-road, driving for what seemed like ten minutes before hitting the main road. Once we were on the paved road, our driver Han floored it, and in no time, we were flying down the road at seventy miles an hour.

I was air guarding again, hanging out that twenty-ton Cadillac was like floating on a magic carpet in the desert; it felt like we would float a little higher every time we hit a bump in the road. Sergeant Rosales was worried that we would roll the vehicle and told Han to slow down, but I knew Han had everything under control, and about five minutes later, we arrived at the scene.

They told us to prepare the fire extinguishers on the way there, and it was quickly evident why. The Stryker was in flames next to a giant crater on the road. All I could see from the Stryker was a massive plume of thick black smoke being blown right towards me. It was a raging fire, and you could hear the pop of rounds going off from the intense heat.

The entire time the Stryker was on fire, I wondered why no one was trying to put it out, but it was likely too dangerous at that point. Suddenly, the wind shifted, taking the smoke with it, revealing the extent of damage the Stryker had taken. It lay there, completely flipped over on its back with the ramp wholly gone, its back end shredded. I was not expecting to see the vehicle's wheels in the air, so I stood frozen for a moment, just looking. The entire backside of the vehicle where I was air guarding a few days prior was mangled because it's where the blast was centered and how the Stryker ended on its top. Some of the guys in our vehicle got off to pull security on foot, and when they did, I pulled out my video camera to record what was going on.

"That's messed up," Timo said when he saw me recording it.

"What?" I said. Playing dumb is mandatory in the Army.

"Recording this," he replied. "How would you like it if I recorded you burning alive?"

"I wouldn't care; I'm recording because I'm going to make a documentary." I had no such intention at the time, but I wanted to record as much as possible since the helmet cam I had Celina send me never made it from the states.

The dismounted troops found a wire leading to the crater, and a few hundred yards away, a man ran into a building. Rather than risk any more American lives, Apache helicopters that came in as support fired rockets into the building the guy ran into. Nothing else happened after that; we sat around pulling security until the medical evacuation helicopter left, and shortly after, we left the area too.

Only one guy was pulled out of the vehicle before it burst into flames, and he got out with minimal injuries. Unfortunately, everyone else in that vehicle lost their lives.

CHAPTER TWENTY-FIVE

THE REAL FUCKERY BEGINS

When we got back to the FOB, I felt lost and empty inside. I wanted to tell Luis about what my patch said; I wanted to hear him crack a joke and say something witty about it, so I headed to his bunk.

Luis had a foam board or something filled with pictures of his family. Seeing all their smiling faces and knowing that they were never going to be able to see him alive again broke me. I knelt next to his bed and those pictures and cried. I cried for all the men we lost that day, so I knelt there for a while.

Things got stupid quick after this incident, with Albie being the highest-ranking person since the commander was gone now. During the six weeks since arriving at the FOB, my buddy and I smoked his entire stash, and during our short three-day mission, I was able to get some hash after asking a few ANA guys. It wasn't much, maybe enough to fill the cigarette one hitter six times while reasonably packed.

I was disappointed that I couldn't get more and felt terrible about contributing to our weed-free situation. When we got back, I brought out the score.

"Check it out, man," I said to him. "Sorry, but this was all I was able to score for us over the three-day mission."

Expecting him to be disappointed along with me, he suddenly jumped up and was ecstatic.

"Bro! Don't trip man; are you kidding me? You're awesome. At least we have something to smoke."

I didn't think about it that way; we ran out right before we left, and I brought back something to help us unwind. It was a great night, as great as it could be, given our situation. Within a few days, we got word of a five-day mission that we would do at the battalion level, and I thought of a competition to have with my smoking buddy to help alleviate our cannabis shortage.

"Hey bud, I bet I'll come back from this five-day mission with more weed than you."

"No way, I'll come back with more."

And just like that, the competition was on. Over the five-day mission, we operated out of the Combat Outpost (or COP), an old ANP compound next to a little market area. We were there for five days, sitting on our asses, mostly. Afghan police would go on the roof at night and smoke their reefer. I have to say that it smelled wonderful, and I began to keep an eye out for particular plants as we passed through the market area.

I decided that I would leave the compound during one of the nights, run down the road, and chop down the few weed plants in the market area to bring back and win the competition. Chester and I volunteered to pull guard from the vehicles that were posted outside the compound walls, and as we walked out to go, I dropped my helmet and body armor and sprinted for the cannabis plants as fast as possible. I dropped gear to be as quick as possible to minimize the chances of one of the vehicles pulling security, spotting me run down the road and lighting me up. To them, I would've just been a human silhouette in a thermal camera, so they would have assumed I was a local up to no good.

Knife in hand but closed, I sprinted to the plants under the moonlight, flipped the blade open, chopped them down, and rushed back with them, in hand, back to the entrance. I cut myself a little when closing the knife for the run, but it wasn't anything significant, and I quickly got the bleeding to stop.

I had a bunch of plants in hand after running fifty yards to return from where I started, and I promptly tossed them to the side of the entrance in the dark. We went to the Stryker to relieve the guys in there, and when they left, I dug through the Stryker to see what I could find to hide the plants in. I ended up finding a trash bag and used it to hide the plants to bring them back for trimming and spent the next hour chopping the primary buds and big leaves in the back of the Stryker.

It took a while to clean up, and when I finished, I stuffed the bud back into the plastic bag and stuffed it into my Camelbak. I ended up with quite a bit, and I had little doubt that I would win the competition.

It all went down the night of the fourth day. We all expected to go back to the FOB the next day, and thankfully we did because I would've had to find a hiding spot somewhere in that compound for my Camelbak had we not left.

A soldier and I previously acquired some dip that the Afghans use, so I asked for some. I quickly regretted it. It made my face go numb, smelled like horse shit, and made me feel nauseous. I didn't like it, so I threw it away, but the other guy seemed to like it.

As soon as we got to the FOB the next day, I saw one of the shit soldiers who had previously gone AWOL get off our vehicle and talk to sergeant Rosales. Then Rosales called over the soldier who enjoyed the dip, reached in his pockets and pulled it out. That shit bag denied ratting him out, but I knew he had something to do with it. He also knew that I tried it, so I wasn't sure what would happen to me. Thankfully, my name stayed out of that incident for the time being anyway.

Back on the mission, I was at the COP compound, enjoying the scent of some Afghan bud. One day, we started a fire at the burn pit on the compound to burn trash, but since there was nothing to do, I volunteered to keep the fire going and just hung out by it. Sitting alone by a fire is something I've enjoyed since grade school—in this instance, sitting there helped me take my mind off the fact that I was on the other side of the world in intense pain. Hanging out around a fire reminded me of all those nights at my ranch in South Texas, no matter the location.

As I sat out there, I noticed a Stryker about forty yards away on the perimeter of the compound for security and that Stryker had that shit bag on it. He got off shift, then came to the fire and proceeded to talk to me.

"I know your secret," he said abruptly.

"What secret?" I asked, playing dumb like always. But I knew what he meant.

"You know, your secret."

"No, I don't know about my supposed secret," I replied. "But since you know, you should inform me."

Then Chester, trying to deescalate the situation, piped up.

"What, that he's a weed connoisseur?"

My stomach flipped. It wasn't a brilliant thing to say to someone who sleeps well at night knowing that he's making people's lives hell out here by being a rat, even as a joke.

"You know, like what that guy who got busted had."

This remark confused me a bit, so I just replied, "Whatever." I wasn't sure how to take that comment and didn't want to overthink it.

The next day, we got back to the FOB, and at night more men were having their rooms switched, so nothing much went on except the platoons changing rooms. Another thing that changed was that as a sniper team, we were no longer a part of "weapons" platoon with the tankers and mortars, and we were transferred to the "headquarters" platoon, which is where we should've been at the very beginning of the deployment.

When I heard we would change rooms because we changed platoons, I was pissed. But since we were getting to where we only had two men in a CHU instead of our three-man sniper team, the hassle was worth it.

Supposedly Albie told Jack, "He takes care of his snipers. You and one of the taker sergeants in one, and Timo and Perez in another."

When Jack told me, he said what I thought. "Wow, this guy is seriously bipolar."

Many people in the Army seem to be, though, and I don't know if the Army does it to them or if they enlist like that. In any case, I'm sure the Army plays a significant role in it.

We moved out of our old CHU and into our new room. The next day was pretty relaxed, even though I had to do an ammo detail, which took about two hours. It was better than being down at the motor pool, though, where all of my old platoon mates were doing a command changeover layout since our company commander was KIA.

After the detail, I went to eat, and as I finished, I walked out of the chow hall. I ran into Randall, the most bipolar person I knew by far. He told us to bring chow to our chain of command because they were the only people at the motor pool and couldn't leave because of the layout.

Tracey (who was on the detail with me) and I went and took the guys on detail their meal, and as we were walking up with one-hundred yards to go, Jack stopped us and told Tracey to take the food to them. After Tracey was out of earshot, Jack told me that he had to search our rooms on Albie's orders because ol' bag of shit told Sergeant Rosales that I had weed. Sergeant Rosales proceeded to skip the entire chain of command and directly tell Albie. According to Jack, that was the story at first, so we went to my room. There, since I had just moved rooms, I had already thrown most of my stuff everywhere to try and to set up my new CHU.

"Good," he said. "It looks like I already made you dump your stuff."

While Jack reported to Albie that he found nothing, I stayed in my room. Not long after he left, Jack came back to my room,

"Albie said you're off the sniper team."

I froze. "What?" I said. "Why?"

"I know it's not fair, but it's what he said. Probably just because your name was brought up."

I was floored at such an abrupt shift in position, but nothing was to be done. The ironic part of it all was that we ended up throwing away all of our cannabis weeks prior because of all the shit that was rolling downhill from a plethora of various investigations into civilian killings in other battalions.

He kicked me off the sniper team because of an allegation, and it infuriated me because I had been a sniper for 90 percent of my military career. I was smoking pot, but so was everyone else, so his behavior was just him being a fuckface for no reason. It was seriously some bullshit. How could Albie claim to take care of his snipers when he believed the word of someone who went AWOL and was pity-promoted to E-4?[14] I put so much blood, sweat, and tears into that company, and within nine months of Albie taking control of the company, it was falling apart at an exponential rate.

When he first got to the company, he had everyone E4 and below in a conference room and told us that we could openly speak our minds about what goes on. Many brought up that most of the leaders are incompetent, and I brought up the fact that Clint and I took an FM 7-8 test along with all the squad leaders, platoon sergeants, and officers in the company. It was just a test of basic infantry knowledge, but I ended up scoring higher than 60 percent of the leaders who took the test, and I was fresh out of basic training. I also brought up that I had been on the sniper team for around two years and had been to a range at most a handful of times. Albie told me that he would change that after cracking a shitty sniper joke, and he also said to us that we weren't special forces and had to work with what we had.

---

[14] E-4 is designation for the rank of "Specialist" in the US Army

Once they told me that I wasn't on the sniper team anymore and that I would be moving to the first platoon, I told the medics that my shoulder was hurting to see if I could get surgery. Rolands, one of our battalion medics, took me to the aid station where I spoke with Captain Callahan while Captain Cleve sat next to him. I told captain Callahan that I had a SLAP tear in my right shoulder and that my shoulder was in pain. Captain Cleve decided to open his big mouth.

"I remember you; I saw you before deployment and called ortho. They said you were good for a year."

It took every ounce of discipline I had to not explode on that lying motherfucker. Captain Callahan thankfully ignored his colleague and asked,

"Okay, you have a SLAP tear. How did you get it?"

"I believe from working out with my platoon. I originally went, and the X-rays said that I had a Hill-Sachs deformity."

"You need surgery," he said quickly. "We are flying you out tomorrow."

What a relief to have a medic who didn't need to google my injury! After he stated that surgery was required, captain Cleve again decided to open his cock sucker.

"Well, can't it wait?"

"No, he needs surgery," Callahan responded quickly.

I sat there wondering what was wrong with people like Cleve. That whole place was corrupt to its core, and during the talk about my shoulder, I wasn't 100 percent sure they were going to fly me out for my surgery. After he tried to change Captain Callahan's mind about me going right in front of me, it was clear that ortho never called him back.

I knew how much of a pain it was to get through because it took me six hours to get ahold of someone from there. I knew he wouldn't spend that much of his own time for me before deployment. It also seemed a little too convenient that ortho said I could do a year without the surgery, which is how long the deployment was. So I wanted clarification.

"So I don't need surgery?"

"No, you do need surgery," Callahan said again.

After talking about my injury, the three of us started to bullshit around, and they asked me about one of our new sniper rifles. They also gave me a mild narcotic, but nothing was going to kill the pain the way cannabis had been doing for me.

They told me to pack and make sure that I took important stuff because I probably wouldn't return. I asked about my personal belongings, and they said they'd send them to me. I left, and after I got back to my room, I sat there for probably a good thirty minutes, just astounded and stoked that I was finally going to get help. It was so surreal that it was happening, and I let all of my friends know what was going on to say my goodbyes that night because most of my friends were leaving the following day on mission.

I debated calling Celina because I knew how that place operated. Something was likely going to happen, and I might get held back. I opted to call her until I was on KAF because I knew that escaping the FOB was the actual task.

While saying my goodbyes, Appleton, my ex platoon leader, talked to me and was sincerely sorry about what transpired. It seemed as if he was crying, or at least on the verge of it as he spoke to me. I thought it was strange that he was being so emotional, but I believe it came from his frustration with how no one utilized the chain of command. He told me that if it had been brought up to him (which it should have), he would have squashed it. I knew he was sincere because he confiscated some hash I scored a few weeks prior that Chester was holding on for me and didn't do anything but scold him. I told him that it was alright, that everything happens for a reason, and that I was just finally glad to get my surgery.

I also talked to Jack, and he let me know that Albie knew I went to the medics and was flying to KAF the next day. That fuckface said I was trying to fly out immediately because the medics sent the request to the chain of command, and they had to approve it. Chester was pissed that I was leaving because we'd been friends for a while, and I was leaving early.

Nighttime came, and I spent most of it with the C. Co. sniper team. Later, I went back to my CHU and packed my assault pack. I went to bed late and was awoken to someone pounding on my door the next morning.

CHAPTER TWENTY-SIX

TIC TAC TOE & GO

I wasn't sure who was knocking, but I opened the door. The sergeant informed me to be at the aid station with my bags at noon for my flight at 1 p.m. I woke up and got ready, then sat around and did nothing with the surreal feeling of "Hey, this is happening."

Noon came quickly; I threw on my rucksack and walked to the aid station. A few people were there, and I asked a captain if this was the wait for the bird to KAF. He said yes, so I dropped my pack and sat on it. Thirty minutes went by, and I was thirty minutes from my flight. Anticipation was building, and I started to experience a slight adrenaline rush.

Then in the distance, I saw Rolands walking up. I had a feeling he was coming for me. I was right.

"Albie talked to Captain Cleve," he said. "You're not going out on this flight."

I didn't overthink this at the time because they told me there was a possibility I wouldn't be on that flight. The request got sent up kind of late and needed to be approved. I grabbed my bags and walked back to my room, bummed out about not leaving. I just laid in my bed, pulled out the laptop I packed, and watched some anime. An hour and a half later, there was a knock on my door. I opened it, and it was a staff sergeant that I didn't know.

"Are you specialist Perez?" he asked.

"Roger."

"Pack your stuff; you're a part of HHC now. Also, Cummings wants to talk to you."

I asked where I could find him, and he told me he would walk me to where he was. We walked to the battalion TOC and waited for him there because he had walked off. Ten minutes later, he showed up, and we talked one-on-one.

"I heard about what happened," he said. "I want you to work for me in the TOC.[15]"

"Well, I still want to get my surgery."

"You can get it when you go on leave. When do you go?"

"Well, when I was in A. Co., it was supposedly in March, but I'm not sure."

"Okay," he replied. "You can get it around your leave time. It's not bad here; you'll be on the radio and relaying messages, twelve hours on and twenty-four off."

It sounded like a sweet deal, but nothing is as they make it out to be, like everything that happens in this place. I agreed to help but didn't have a choice. I planned to talk to the medics about my surgery and tell them that I wanted to go out around November. The only reason I didn't go and speak to the medics again right away and helped Cummings was that I respected him. He was the first real first sergeant for Alpha company and was a damn good one. Morale in the company was very high when he

---

[15] Tactical Operations Command (TOC) is the tent that a battalion operates through for battlefield operations.

was in charge, and all the other companies would get in trouble just about every weekend.

That day, he said, "Albie gave up on you, Perez. I never did, so don't give up on me."

The gig was a good look into the behind-the-scenes of the battalion. However, the job was also different from what they said. It was not twelve hours on, twenty-four off because there seemed to be no rules around telling lies—the middle of nowhere is where mistruths fester.

Another three days went by, and while I was at work, I ran out of my pain meds. I went back to the medics at lunch to get some more and talk to them about getting my shoulder surgery. I was determined to fly out the first week in October. I wanted to help out until November, but working with these fucks drove me insane. I walked in, and captain Callahan was there,

"Hello, sir," I said. "I'd like some more meds for my shoulder and would like to talk to you about my shoulder surgery."

"You're not getting it anymore."

"What?" I cried.

"Come back at 1600, and we will talk about it with captain Cleve."

Infuriated, I went back to work and told Eddie what they said. I also told him I would talk to them about the surgery and let him know what they said. I knew he believed me when I told him that my shoulder hurt because he knew what kind of soldier I was. I never bitched unless I had a reason to, and lately, that was all that I had been given: reasons to bitch.

There wasn't much Eddie could do as a staff sergeant—only the bipolar, lying, cheating, corrupt ones who had fifteen-plus years in service got to run and ruin our lives. All he did was pass the word along.

1600 came around, and I sat in the aid station for a bit, waiting for captain Cleve to come back from whatever he was doing. He showed up and started our conversation. But this time, I had my voice recorder ready. Below is the transcript, in all its glory! Not listened to by anyone for a decade until this book! I wanted to document these fucks lying about all this.

Callahan: We've had a recommendation, per ortho, and we've made the recommendation that you can deploy and do your job with whatever issue you have. What have you done for rehabilitation while you've been here?

Me: Here at Frontenac?

Callahan: Yeah.

Me: I was out on mission sir; I mean I was actually working.

Callahan: Oh. Reason, we make recommendations to commanders and first sergeants and based on first sergeants' decision to keep you here if your soldiering is suspect and they're kind of worried that you're using secondary gain to get out of here.

Me: Well that's bullshit sir, because it's been hurting since March.

Callahan: I understand that.

Me: And I'm not even in A. Co. anymore

Callahan: I understand that,now again we make recommendations. So while we are here, we have plenty of time to do rehab. I know how to do it, I can tell you how to do it, and we can monitor. As for getting you out of here it's on the command, we've made our recommendation.

Me: But it doesn't make a difference If there's new command that I'm under now? I mean can't you just shoot up the recommendation to them, because they seem pretty adamant about getting me out of here.

Callahan: Who's your first sergeant?

Me: Staff Sergeant Eddie.

Cleve: What company are you in now?

Me: HHC.

Cleve: HHC, we'll talk to the command.

Callahan: We'll put a recommendation out, it's not going to change anything. Again, we recommend stuff, we don't make stuff happen.

Me: I mean you just seemed pretty adamant that I needed the surgery sir, and I mean it does hurt and it's hurt for a long time.

Callahan: Then there's an orthopedic surgeon?

Me: Was that on the SLAP tear or on the Hill Sachs deformity?

Callahan: Uh, there were others.

Me: He looked at both of them?

Callahan: I didn't talk to him but . . .

Cleve: Yeah I talked to them. I mean, we talked to them together.

Me: Yeah, but that was about the SLAP tear sir. You didn't know about the Hills Sachs, because that was the X-ray, I had you look at the MRI.

Cleve: Yeah, but they look at everything. You know what happened was I called them [smack], and uh and they went to look at all the films, and it was an ortho PA and they took down the information, they talked to their supervising staff and then and then called me back. So they had a chance to look at all the radiographic imaging. We can talk to your command but you know at the same time we're uh, the prior alpha command was adamant about you at least giving a try to do short patients here. You can work with us in terms of . . .

Me: Well, I mean I was on profile for like two months.

Callahan: Listen, don't interrupt him.

Me: Oh, I thought he was done, sir.

Cleve: I mean your last chain of command wanted you to at least have a try here of limited duty, working in the TOC, and seeing what you can do, and I was fine with that. That's fine, I didn't realize you moved over to HHC, we can talk to them and see what they think. If they're adamant about getting you out and you know and the surgery then we can start working that. Last time first sergeant was adamant that we at least give you a trial of physical therapy here which, uh, you know is fine. I just didn't want you out there re-injuring it. Um, so uh your first sergeant now is, who, first sergeant Reed?

Me: Ummm, I fall under MSG Cummings.

Cleve: You fall under MSG Cummings, cause he's in S3.

Me: Yeah that's where they moved me, S3.

Cleve: Okay,[inaudible] he's very capable at overseeing a rehab program while you're here in theater. The orthopedic surgeon said, yeah you're eventually gonna need surgery umm and that's the truth. But they also said it can wait a year while you're deployed. It doesn't sound like you're back doing the job of, uh, um, that would [inaudible] in the first place so. Based on how many people we have outta here we need every person here we can get.

Me: I understand that sir, but y'all put me on profile for two months, and the major pain subsided but like I said it was. Certain areas that still hurt. So I mean, even with the rehab it might make the constant pain stop, but the actual it still being hurt is still there.

Cleve: I mean what are you doing now, you're working in the TOC?

Me: I'm working in the S3.

Cleve: That means, well is there [inaudible] physical activity that exacerbates.

Me: Well, there's times that they say they want to send me out as an air guard and just you know carrying stuff I do still carry heavy stuff for them.

Cleve: we can talk to them to see what they think. I mean like I said, you know it was initially that the command wanted you to uh wanted you to to uh you know or for administrative reasons and I said well that's fine. I think we can work within the profile, we can work within the diagnosis to make that happen as long as they didn't have you out doing the same kinds of things being a sniper or things to exacerbate it. So you know if if the command now is adamant about getting you out, that's fine with me. But if they have a job for you here and need to keep you here, we can talk to them and say look these are things they may not have you do, but, you know orthopedic surgeon said this can wait up to a year if we're not doing anything that is going to exacerbate your shoulder.

Me: Okay.

Even more enraged than I was at the beginning of the day, I went back to the TOC and told Eddie what they said again. A few days passed, and I continually tried to get them to believe me when I said that my shoulder hurt. It seemed like only Eddie believed me, and still, no request got sent up, not that I'm aware of anyway. I was beyond frustrated since all I did for that place was bust my ass and work hard. So hard that I injured my shoulder and didn't get anything back but was lied to and shunned. Any job that puts your health in another person's hands is not a job worth having.

The worst part of this whole situation was I lost my belief in merit. Before all this, being a good soldier meant something. Now, I knew the only way to get off the FOB was to start playing chess with these pea-brained idiots.

CHAPTER TWENTY-SEVEN

ALMOST DUSTWUN

A few days after going to the medics, O'Brien brought me a paper for our R&R.[16] It was an Excel worksheet with dates and names, and he asked me which date I wanted. I saw an open slot in November and thought, "If they don't want to help me, then I'll help myself on leave." I put my name down for November and all around stopped bothering them about my surgery. I knew they wouldn't help me, and a couple of days later, I went to Cummings.

"Hey sergeant," I said. "Remember how you said that you'd take care of me and let me get my surgery around leave?"

"Yeah."

"Well, it turns out that I go on leave in November."

"Okay, yeah," he said, interrupting me before I could continue. "I don't see that happening; I'll talk to Sergeant Major Driver though."

I honestly couldn't figure out their problem; everyone acted like the unit would fall apart if they sent me for surgery. Like Specialist Perez needed to be in the TOC, behind a desk on the radio and laptop to keep everything together. I know he didn't talk to the sergeant major about it, and if he did, I know that he would never let me go. Infantry idiots convinced the medics who were so adamant about me getting

---

[16] Rest and Recuperation (R&R) is military slang for the free time of a soldier. Typically 2 weeks for a year deployment.

surgery that I was a shitbag and told the entire chain of command that it's "an elective surgery and that it can wait."

I think it was a mistake because a few nights prior, I went to S1, the people who deal with paperwork, to find out exactly what day I was going on leave in November. I wanted to know because I had already told my mom that I was going. She told me to try and get an exact date because if it wasn't around Thanksgiving, she could still get things together to celebrate the holiday while I was in town.

The people at S1 looked it up on the computer, and my name wasn't on there for November, nor were there any slots open. I asked him to look up October, and there were three slots available for October 26th. I asked to put my name down for that date, and they told me the first sergeant had to approve it first, so the following day I brought it up to Eddie, and he brought it up to O'Brien. I was off that day and used it to get some rest and watch a few movies, but back on shift the following day, I asked Eddie what O'Brien said, and he told me to talk to him myself, so I did.

"O'Brien," I said. "Remember that leave signup sheet? I signed up for November but looked it up, and there weren't any spots open, and my name wasn't down, but there were some open spots for October 26th. I was wondering if I could go then."

"Well," he said. "Your surgery is elective . . . "

After those words hit my ear, I stopped paying attention for a bit. O'Brien's final words to me were, "Cummings and I talked about it, and now your leave is in April because we think that you'll go and get surgery on leave and won't come back."

They put me on what was probably the last leave flight from there because they were afraid of me getting my surgery if I left soon. They should've been more worried about me shooting them in the face for being fuckbags!

I tried to keep my cool and did so for about fifteen minutes, and after he told me that, all I said was "Roger sergeant." That's all that I could do in the meantime. Trying to argue with a sergeant who couldn't even tell me how many weeks were in a year was a waste of my breath. I sat down, head in my hands, when the guy on shift asked,

"When do you go on leave, Perez?"

"I don't know."

"Really? I heard they were screwing you over and putting you on the last flight out of here because of your shoulder surgery."

Now, hearing that from an outsider pissed me the fuck off. I ground my teeth and was so frustrated that my vision started to blur, but I kept doing my work like a good soldier.

A few minutes later, I was asked by the "battle captain" to find some guys and get info from them about their flight. I said Roger, got up, stood next to him, and got a notepad for him to write what he wanted me to ask them. When I did that, I got in front of someone as they walked into our tent section but wasn't sure who it was. I had only seen them from my peripheral vision. I didn't even cut in front of him, but he began to bitch at me.

"Oh, it's okay," he said. "I wasn't going to say anything important; it doesn't matter; I'm no one." Guess what rank he was!

"My bad, first sergeant," I said. "I didn't see you there," I said this even though I did see him. I didn't know his rank or if he wanted anything from the captain.

"You know what?" He shouted. "Go stand in that corner and stare at the fucking wall."

I turned toward him, got in his face, and yelled, "ROGER SERGEANT!" and walked out of the tent. I was beyond done with those incompetent fucks; they had done nothing for me the entire time except make this deployment the worst experience of my life.

Not only were my friends dying and getting hurt, but I had to deal with self-centered idiotic bitches who were taking advantage of us being in the middle of nowhere, without a way to stop the abuse. I decided then that I wouldn't do anything for that place anymore. I would slack off on shifts, refuse to help out, and close my ears to their orders. They wanted to treat me like a bad soldier, so I would act like one.

I went to my CHU real quick to assess the situation. I was thinking about leaving the FOB on my own and seeing how far out I could get. I quickly realized that would lead to my or my brothers' demise. Then I thought about posting up right outside the FOB walls and digging myself a hole to hide in for as many days as possible to make them worry, but eventually decided to head over to Clint's CHU to cool down and hideout for the meantime. I figured they were too dumb to think about looking there, so I hung out with him and wrote a lot in my green book at this time, most of which make up this book.

We talked for a few minutes, and then he crashed out because he was tired from just returning from being outside the wire. I decided that I would call IG hoping that they would do something to help me. They were my last chance to get my surgery done and keep my sanity, and I thought that if they were to fall through, then the military was indeed one of the most corrupt organizations ever. They claim a lot and talk big but don't back it up.

IG is supposed to help you when your chain of command doesn't help or listen. You might think that you could never kill or hate anyone, but if you were unlucky enough to get in a unit like this, you would find it easy to hate. I could gut some of these fuckers like a pig and gladly bathe in their blood.

Speaking of death and bad luck, on September 26th, we lost another good man from our platoon: Specialist Adam, a mortar. His vehicle hit an IED, and he was the only one killed. Everyone in the vehicle avoided severe injury because the IED just blew up the driver's hole. When we first arrived at Frontenac, the 1/17 Infantry Buffaloes lost nine men in two weeks; the unit took a severe hit. Alpha company had almost gone a month without losing a soldier.

Sadly, more good men were losing their lives. In the meantime, the ones who made hell for us sat there on the FOB, continuing to breathe. Those eight hours of hiding out gave me plenty of time to write and get a lot of stuff off my chest. I was on the verge of doing something much worse.

As I calmed down, I thought, *I'll go to work tomorrow only because I want to be one of the first to know when we can use the phones. That way, I can call IG.* Whenever a soldier dies, we go on "commo blackout." That way, the Army can inform the family of the deceased that their loved one is gone before the dumbasses of the unit call their wives and tell them, which propagates distasteful and disrespectful rumors.

Around 7 p.m., we left Clint's CHU because he had awakened from his slumber and wanted to eat. I decided to go and talk to Jack because another sergeant told me he was looking for me and had heard a rumor that I got on the flight to KAF. I went by, and Eddie greeted me as soon as I opened the door.

"Where have you been?" he asked me.

"Hanging out in the CHUs," I said.

"We've been looking for you all day," he said, looking concerned. "Everyone. The battalion commander, sergeant major, captain. So, what happened earlier?"

I told him what happened with the first sergeant and me.

"Okay," he responded. "So what are you doing here?"

"I want to talk to Jack."

"Okay, well when you're done, go to the TOC."

"I'm going to have shift tonight?" I asked this as if I didn't expect that as a minimum for being gone the entire day.

"Yes, since you were gone all day, you're pulling shift tonight."

"Roger, Sergeant." He left me behind in Jack's CHU.

I saw Jack hanging out in the CHU, doing God knows what.

"Hey Jack," I said. "What did they say?"

"Nothing much," he said. "Just that they were just looking for you."

"Good," I grumbled, crossing my arms. "I'm glad they were worried. That's what they get."

"Yeah, but Eddie wasn't outraged after I talked to him."

"What did you say?" I asked sincerely.

"I just told him that you did it because you are frustrated and in pain and that you want your surgery. What he just told you was just him being a staff sergeant."

"Well, I don't care; I'm fed up with this place." I plopped myself down next to him, and we started bullshitting, as we always do.

CHAPTER TWENTY-EIGHT

SURREAL INTERACTIONS

After that, Jack would walk to the TOC to check if the commo blackout had lifted yet. He asked if I was going up there immediately since they ordered me to.

"No," I responded. "I'm going to my CHU real quick to get my headlamp."

"Okay, see you later," he said, and we went our separate ways. But on the way to my CHU, I realized I had my headlamp on me already. When I got there, I just sat down and talked with my roommate, Landen. He told me that many people had stopped by looking for me, and I told him that I stopped by Jack's CHU and was updated on the situation. I walked into the TOC and standing there was O'Brien. I sat, ate a small bag of chips, grabbed a bottle of water, and began my journey into the unknown.

"Where have you been?" He demanded.

"Just taking a break."

Suddenly, Cummings walked by us. "Take a break outside," he said.

"Roger, sergeant."

I put my headlamp on as I walked out, and we walked fifty feet from the TOC. We moved so far away because the generators for the tent were extremely loud, and even at a distance we walked, they still created communication issues. I even had my voice recorder ready in case I needed to capture even more evidence of their bullshit, but all you can hear on the recording is the generators. While we were walking, Cummings was looking around somewhat erratically, as if he was trying to find a place in the distance to make me run towards. Suddenly he stopped and pushed me in the shoulder.

"WHAT THE FUCK?!" he shouted in my face.

"I needed a break, master sergeant," I said. I didn't want to do or say something that I would regret or get me into more trouble.

"Don't you think you should've at least told someone where you were?" he continued. "Do you know where we are?"

"Afghanistan, master sergeant."

"You're right," he said. "In a *motherfucking combat zone*!"

"I didn't mean to make you guys worry, master sergeant," I lied.

"What's your problem Perez?" He got closer to my face. "You acting up this way because we took away your leave?"

"That's part of it," I explained. "It's because before we deployed, when I brought up my MRI results to Newt, he brought it up to Albie, and he said that I could get my surgery once I was in Afghanistan."

"What did I tell you?" He cried. "I told you that I would take care of you. I don't care what rank you are; I treat everybody like a man. Your rank is simply where you stand in the Army. I didn't expect you to do something like this, Perez."

"Roger, master sergeant," I said. I thought I was getting somewhere with Cummings. "I just didn't want things to be like this, but I'm fed up. My shoulder legitimately hurts."

"If you go on leave, can you guarantee me 100 percent that you'll come back?"

Now, that was a tricky question. As usual, I was a dummy and did the right thing: I told the truth.

"Probably not, sir."

Why did I tell the truth again? Because a lifetime of experience taught me that it was likely the best course of action given the current predicament I found myself in. Telling the truth in dire situations has benefited me in the past, so regretfully, I did so.

Cummings stepped back and sighed. "You know what, Perez? I'll let you go to KAF to get a new MRI done."

I looked at him, shocked. "Seriously? That's awesome!" I cried. "That's all I want: to be helped. You scratch my back, and I'll scratch yours."

"What else do you need done?" Cummings asked.

I thought about it for a moment. "I'd also like to get X-rays done because the X-rays showed the Hill Sachs deformity."

"Okay," he said. "If they say that you need surgery, then I'll let you go, but if they say you don't, then you're staying here and not going on leave anytime soon."

"That's fine," I said quickly. I didn't want him to back out of the deal now.

And that was it. Cummings also told me that I would be getting an Article 15[17] because I left my place of duty and that the battalion commander and sergeant major would ream my ass.

---

[17] *Article 15s* are a mechanism that allow the chain of command to punish a Soldier for offenses under the UCMJ without formally charging him/her at a court-martial.

After that little chat, I went back to the TOC and waited around for my shift to start. Another Captain from the unit approached me, and he began trying to tell me to talk to O'Brien.

"I already did, sir," I responded. "I also talked with Cummings too. It's been taken care of."

"Okay," he stammered. "Well, there's going to be more to follow."

I scoffed. I wanted so badly to yell, "Yeah, okay, FUCKFACE!"

I waited around for the shift change brief to finish, but before they started, I think that Eddie, O'Brien, Cummings, Morton, and Driver all had a pow-wow about what to do with me and what I did. Nothing happened, and I'd be damned if they would've done anything. Nothing happened except for me having to stay awake another twelve-plus hours. They told me to get a haircut during that time, but that's nothing new. My entire military career, except basic, I've pushed the standards of hair acceptance.

The following day during shift change, which happened every twelve hours, the brief included the phrase, "Expect to hit IEDs," which meant we should expect to get blown up at any moment.

Wow, I thought, but it was the truth. *You can't do a massive mission in this country and expect not to hit IED's*. I could only hope that no one got hurt.

On October 1st, 2009, I sent IG an email. I believe it was the first time I attempted contacting them. It took me a while to write the email because I gave them all the information on everything that was going on. I also told them that I didn't want them to get involved unless they could help me.

If IG got involved and didn't get me out of there, everyone was going to be pissed and make my life even more hellish than they were already doing. I was hoping some good would have come from emailing them.

But I didn't have time to think about it much because the mission had begun. I hoped everyone would be alive when I got back on shift fourteen hours later. We changed our frequencies to our radios for communication security reasons and, in doing so, created problems with the network and made it difficult to communicate with each other.

On October 2nd, IG responded to my email. The people I thought were supposed to help soldiers in need when no one else would turned out to be another part of the machine. They said that they are a "staff" organization and that all they can do is talk to the chain of command. Therefore, they can't guarantee that they could help me get my surgery. They recommended that I use the "open door" policy and talk to one of my commanders; those were the fuckheads who didn't want anyone to leave! They also recommended that I speak to the chaplain or a stress management team, but little did they know that the chaplain was just as corrupt as everyone else here. When I read their formally worded refusal for help, I wanted to flip over my laptop and trudge out of there on my own.

Fuck it, I thought. *No one is going to help me. Well, that's fine. If I get the X-rays and MRIs and they say that I don't need surgery right away, that's fine too. Fuck it. Fuck it all.*

All the shit that they and the Army were doing to me was just fuel for the fire. It's what led me to keep writing. I was determined to document everything and write this book one day, and here I am.

I wanted the world to know exactly how the Army—particularly this unit—treated its soldiers and how no one was willing to help. I was glad it was all happening, in a disturbing sense, because I felt like they would all pay in the end. They won't ever deny and claim ignorance about my situation because *everyone* knows my situation. When they couldn't find me for eight hours, they had everyone looking for me. Everyone knew that the worst offenders in charge of the unit, Morton and Driver, were going to "ream my ass," but they never did.

"I have the worst luck," I said to myself that evening. Why did all this happen to me?

All I ever did for the Army and that unit was good; why did they mistreat me? What have I done to deserve this?

My train of thought was an attempt to maintain my sanity. I believe that everything happens for a reason, so maybe it was happening to me so that I could let the world know the truth. Everyone hated the Army because of that unit, and it was mind-blowing that a unit built from the ground up could be so broken.

After a while, I debated emailing a member of Congress about what was going on but didn't. If it weren't for Albie, I would've likely already had surgery, and the thought of that made me want to cut his head off. I hated thinking like that but couldn't help it. All the physical pain and mental torture were becoming too much, and I knew I wasn't going to be able to take it much longer. I was losing my patience and my mind, and at that point, I made up my mind that when I got to KAF to redo my X-rays and MRIs, I wouldn't go back to Frontenac.

My boots would've hit KAF, and there would've been nothing they could've done to get me back to Frontenac because I wasn't helping people who didn't care if I lived or died. I planned to stay on KAF to support the brigade and decided that I and the battalion had to part ways for everyone's well-being. Especially with Morton in charge. The other day, he came to us in the TOC and told us to fix his weapon because his laser wasn't working. They tasked me to fix it because "I was the sniper," so apparently, that qualified me as an armorer.

The first thing I did was change the battery, and that's all it took to "fix" his weapon. A lieutenant colonel in charge of a battalion of men and been in the Army Infantry for probably fifteen-plus years didn't think to change his battery.

As I handed it back to him, he pointed out his weapons number, a number assigned to us that allows us to determine whose weapon is whose. Morton's number was one, and he didn't hesitate to point it out to me.

"My weapons number is number one because *I'm* number one," he said. I don't even remember how I responded to that ridiculous statement.

He did another crazy thing when I worked about two weeks before that incident. For my task that day, I was in the TOC behind the radio, which was quiet and boring as hell. I was looking at the new Corvette in a Men's Health magazine to pass the time. Morton walked by and asked what it was, so I told him it was the new Corvette. He began to say to me a short story about when he was a young lieutenant and how his roommate had two Corvettes.

I didn't pay much attention to the statement or give it much thought. A few hours later, he walked by again. I moved on to doing a crossword puzzle on my iPhone and didn't hide that I was doing it because I figured that reading a magazine and doing a puzzle on my iPhone was no different. Well, using reason wasn't something that Morton was particularly good at.

"You better put that away," he told me as he stood behind me. "If I catch you with it again, I'll break it over your head. How does that sound?"

"It sounds good, sir!" I replied loudly. I began to put it away while looking at him dead in the eyes. The look on his face was priceless. He was caught off guard by someone looking him in the eye and throwing his bullshit right back at him.

That was the last time we ever exchanged words, and I believe that was part of the reason why he or the sergeant major never "reamed" my ass. They knew I'd had it with their shit.

CHAPTER TWENTY-NINE

ALMOST AT THE END

By this point, I had traded quite a few emails with IG, and the lady I was in contact with reached out. She wondered if they'd scheduled my MRI and X-rays. I wrote to her previously to let her know I couldn't talk to anyone in my chain of command because they were the ones that were screwing me over. The chaplain couldn't help me because he was just as corrupt as the others, and I told her that I would take care of it myself.

She reached out with concern about how I was going to take care of it, and I told her that they hadn't scheduled me because we were in the middle of a military operation and that I would give them a week or two after it to get me to KAF. I also told her that I was not coming back to Frontenac once I was at KAF, regardless of the test results. I explained that I didn't care how much trouble I got into because I didn't plan on reenlisting, and she wrote back asking if it would be okay for her to contact the brigade command to represent me and tell them my situation.

I wanted to say yes, but I didn't because the brigade would call these fucks and ask them what the hell was going on. Then these fucks were going to lie through their teeth to save their asses, and then hell would be brought down on me. I sure as hell wanted someone to help me; I just didn't know what to do.

As anyone could see, nothing was working out in my favor. I wanted to talk to someone, but most everyone was out on a mission, and when I went to work that evening, we lost another man to a road-side bomb, which affected the camp's mood.

While waiting for the mission to be over, I couldn't help but think about the different ways anyone could get away with killing someone out there. It wasn't a normal way to think, and I knew it at the time, but what scared me the most was the thought that it was possible to formulate a perfect plan and attempt to execute it.

Then, something happened at work that allowed me to break away from this train of thought. That day, the battle captain talked about his best army experiences, and I started thinking about how my contract obligation was up in fifty-three weeks, and I'd be off active duty. The idea of next year was all I could think about, driving off-post one last time and putting this miserable hell behind me. Thinking about going to college for free on the GI Bill was also another thing I was looking forward to because I sure as hell didn't plan on working for anyone ever again if I could help it.

Another thing I kept wondering about was why everyone wanted me to stay on Frontenac so badly. They initially told me they didn't want me to get my shoulder surgery because they needed me to help in the TOC. They didn't want me to go because they needed me there, setting high scores on iPhone games, watching movies, getting on the internet, and talking with Celina an hour at a time most days. They didn't fucking need me there; I pulled a twenty-four-hour shift because I disappeared, and they did just fine. I felt like my responsibilities in the company were complete horseshit: anyone with gray matter between their ears could've worked that radio, and the more I wrote about it all, the angrier it made me.

I thought about calling ortho and talking to them directly, thinking I could catch Cleve in a lie. He had told me that I needed surgery, but it could be put off for a year. Such a statement sure seemed too coincidental. I knew he was lying, and I was trying to catch him in it; I wanted the truth to be known. I had to wait to call them over the weekend, though, and I figured it would've been easier to have Celina call them for me, so I sent her an email with what I wanted her to say.

She called on Monday, and the person who answered was unhelpful, but Celina remembered she had a friend who worked at the hospital and was going to get in contact with them. At the end of my shift that day, I left during the shift change brief to go and use the restroom.

Usually, what would happen was as soon as one of the officers gave their final brief, we would get out of there quickly without having to obtain permission to do so from an adult. Eddie told us before to stay for the brief, which I usually did, but that day I had to piss, so I didn't think it would be that big of a deal if I left since the shift change brief was about what happened during my shift, so I wouldn't be learning anything.

The relief had already been informed about the nothing that happened during shift and took over radio duty, and as soon as it ended, Eddie asked Landen where I went. Landen told him that I went to the restroom, and Eddie scoffed.

"Bullshit," he said. "You've got two minutes to go and find him."

After using the restroom, I went to grab my laundry. As I said, the brief had likely ended, and I thought it pointless to go back. As I walked back to the CHU, laundry in hand, Landen and one of the unit's staff sergeants turned the corner.

"Dude," Landen said. "Eddie wants you back in the TOC."

I muttered under my breath. I was so sick of getting in trouble for stupid shit and incredibly sick of their fucking games. I walked back to the TOC, and as I was walking up, Eddie walked out.

"Where'd you go?" He demanded.

"To take a shit, sergeant," even though all I did was pee and did other things at that time.

"It couldn't wait ten minutes?"

"I had the bubble guts from breakfast."

He looked at me like I was full of shit, even though I just told him I emptied myself, and he told me to go inside and talk to the sergeant on duty.

Shit, I thought. *They're going to make me pull the day shift too.* I walked inside the TOC.

"What's up, sergeant?" I said. "Eddie told me to come and talk to you."

"I have to type up and print out your counseling,"[18] he replied.

I said okay and waited around for about an hour and a half before he finally got done. He typed up three for me to sign, one for me leaving to use the restroom, one for me leaving the other day when they couldn't find me, and the last one was one stating that they weren't going to promote me, even though they told me they planned on promoting me to sergeant before I took my eight-hour hiatus. Instead of sending someone to go and get me after he finished typing them, the bitch asses had me wait after a twelve-hour shift while he worked on them.

Counselings were the army's way of telling someone what they did wrong and what they could improve. You can either agree or disagree with the statements, and they provide space for you to give the reasons why you think you don't deserve the negative counseling.

---

[18] Counseling is the process used by leaders to review with a subordinate the subordinate's demonstrated performance and potential.

Counseling can also be a good thing for the record, but those were rare, even the yearly mandatory ones. The first one I spent time on was the one that talked about them not promoting me to sergeant; it said: "I lack the responsibility and professionalism that the Army looks for in an NCO." All three documents had maybe five sentences each, and I believed that Eddie told him to make me wait and purposely go slow because I hope he wasn't that slow for that sergeant's sake.

I took the document and began to write. "I disagree," I wrote, "Because if an NCO is supposed to be responsible, he has a responsibility of taking care of his soldiers. I have not been taken care of; I got treated like a bad soldier because I complained about my shoulder pain. I don't know if you think you are helping me by keeping me here, but you really aren't. Both my physical and mental health are suffering because you don't want to help me."

For the shift change incident, I wrote, "I disagree because I had to use the restroom during the brief. I already briefed the relief and they took over shift responsibilities. The day shift change brief explains what happened while I was on shift and would agree if it would've been the night shift brief as that is the one that informs me of what happened while I was off shift. Also, Eddie never told us to stay after today but we have been told to stay for the brief I just had to use the restroom. We've been told in the past to stay for the brief, not that we needed permission to be released. There were many times in the past when I would stay for the brief and as soon as it was done, I would walk out without permission and never get counseled."

Finally, the one about me leaving for eight hours that day without letting anyone know where I was said, "I disagree because if I hadn't left the way I did, I probably would've done something way worse. I was never sent to get the battle roster numbers though that was the task I was about to be given before I left. You speak of me having a lack of responsibility, and as I have said before it is your responsibility to take care of hurt soldiers. I'm sorry that I didn't get the injury from a gunshot wound or IED and I tried to get it taken care of in garrison, but it was the lack of responsibility in my chain of command that kept me from getting it fixed. Is that the kind of responsibility the Army is looking for? Well I'm glad that I'm not getting my E5 (sergeant) then because as an NCO I would never treat one of my soldiers like that. I was part of a sniper team but got kicked out because of the great NCO the Army chose as a first sergeant for complaining about my shoulder pain. There are soldiers in the unit that have got DUI's, article 15's, and failed PT tests, but this place has no problem promoting them to E5. I've never done any of that but just because I hurt my shoulder soldiering hard for this unit I get cast aside."

After I finished signing them, the NCO who was typing them read them and tried to help me out. He asked if I talked to anyone, and I told him that I'd spoken to everyone, and no one was going to help me. Given my prior interaction with him, he said I should see the chaplain, who wouldn't have helped either. Anything that would've left his mouth would've just pissed me off further. He asked if my MRI and x-rays were in my medical file, and I told him they should be but that the MRI may not, as I got the results the day before deployment.

I assured him there was nothing he could do to help me but was thankful that he at least pretended to care. After that, I left.

Back to my room, I told Landen what happened; we would usually lay in our beds after work and talk about how pointless everything we did was and how we couldn't wait to get out of the military. The next day I was looking forward to running into Eddie and seeing what he had to say about what I had written in the counseling. I hoped that they would get me in trouble for speaking my mind. They didn't bring anything up, and during the shift, I asked Eddie if he could figure out when I would be going to KAF.

I thought I heard them say something about getting me to KAF once we could get a rotation for the radio shift, and later that night, Eddie told me they were trying to get me a flight out for the eleventh. It was the ninth of October in the evening, and there were no flights planned for the following day. I figured they told me that to keep me quiet for the next few days. It was that, or they were trying to line up a 5th brigade healthcare provider, and there was no way in hell I was going to see anyone from the brigade since they could've easily just lied like everyone else in charge of that unit.

The day before, the staff sergeant with Landen prior was late for his shift; he slept through the entire shift change brief to inform him about what went on and what to expect. Did he get in trouble or get negative counseling? Of course not—he was part of the club, and the double standard that was rampant in that unit was what drove the lower enlisted insane. I didn't care because I liked him, but I'm just pointing out the obvious.

After the brief, Morton left his little notepad next to the other guy on the radio. Once he got off the phone, he reached over, grabbed it, and hit the radio guy on the head on purpose with it. It wasn't hard or anything, and if it had ended there, I would've never written this story. It was what happened next that completely blew me away and bothered me.

"If I hit you, it's discipline. If you hit me, though, it's an Article 15!" He sat back in his chair and put his hands behind his head. "Ah, it's good to be the king!"

Seriously? This guy thought he was untouchable, and I believe things like that are what made his subordinates dislike him so much. Then, like always, they fucked me over again.

When I got on shift that day on the 11th, they told me that I wasn't flying out to KAF because they didn't have MRI or X-ray machines there. They told me they would call Fort Lewis and have my medical records sent for their medics to look them over. At that point, it didn't matter because they were going to lie no matter what showed on my records.

I was still going to talk to Cummings to tell him that I wanted to go to KAF for a second opinion and was also planning on calling ortho and emailing IG to let that lady know what was going on. After weeks of calling, I finally got through to ortho, but like everything the Army runs, it was ass-backward, and they weren't willing to help me. They told me they couldn't help me because I'd never been to their clinic.

They told me to talk to the battalion medics because they were the only ones who could recommend anything. It was as fucked up a situation as it could've been when I look at it in retrospect. The battalion medics whose careers were utterly dependent on the Infantry leadership would do whatever that leadership told them, so their careers would be unbothered. And some, like Cleve, seemed to embrace making people's lives hell to do whatever it took to keep people in the unit and on the FOB.

Simply going to them was the fucking problem; they were the fuck heads who didn't want to help me. How could they have expected me to go into ortho if I got the MRIs the day before deployment? Nothing made sense.

This was when I started reaching for the most out-of-the-box solutions that I could think of. I thought about calling ortho back and posing as Cleve but didn't go through with that. Then I thought of having Celina call the hospital commander or contact a state representative. Someone had to want to help me, and I was trying to think of anyone who had the authority to do something.

If I weren't mentally strong, I would have likely already done something at that point. My eye was beginning to twitch from the months of stress I had been under, and it was annoying as fuck. People snapped and assaulted each other in the unit over circumstances much better than the hell they were purposely putting me through.

I was starting to think about shooting myself at first, but then I just thought about shooting myself in my shoulder. Suicide was on my mind a lot, though, and I planned on sending letters to different places, telling the story I tell now. I thought about death a lot at that time, mine and theirs, and looking back on it, the thoughts I had were nothing to do with me and who I was. It was my circumstances and intense undiagnosed mental illness that brought me to some of the darkest thoughts I'd ever had in my life. I was thinking harder about how anyone in this camp could get away with anonymously slaughtering as many in the command as they could. It was sad to be in a warzone and want to kill the people in the unit more than the "bad guys" we were supposed to fight.

CHAPTER THIRTY

THEY WON THE BATTLES, I WON THE WAR

On the 17th of October, I was reminded that I wasn't going to KAF and that the medics received my medical records. Apparently, I could do without surgery.

Well, of course, they were going to say that; they already told me I was good to go without surgery for a year. Eddie seemed to think that rehab would help, and he said to me that Cummings would let me take leave whenever I wanted if I promised not to get my surgery. I promised them, but in the same way they promised me all kinds of things over the last few weeks.

I asked if I could go on October 27th since HHC had one more spot available. I was hoping they would let me go since I didn't want to wait until December.

In the end, though, it didn't matter. I didn't think I was ever going to get out of there. However, something brought me a little hope: another soldier who was on Xanax for anxiety was being flown back to the states because they couldn't get any out there. He told them he would call a senator that was a close family friend. As soon as he mentioned it, they quickly took action and got him a flight home. The idea of an influential person doing a good thing for a brother of mine gave me hope. Maybe if I contacted someone like that, they could do the same for me. The fact that I had conversations recorded with them blatantly lying to me was all the proof I needed.

The days passed, I tried to get on a flight, and a few days later, I was told my request for the 27th got denied because it was close to the date of my inquiry. I knew that I was close to making things happen; I could see the checkmate coming together and was looking forward to putting my plan in action. I was going to call ortho and try and get someone with a heart or conscience, but that was going to take some time. The days passed, I tried to get on a flight, and a few days later, I was told my request for the 27th got denied because it was close to the date of my inquiry.

I talked to the people in charge of leave that morning to see if it was true, and of course, it wasn't. They said it didn't take much to get someone out on leave, so I brought it up and talked to O'Brien about it. He said some bullshit about it being easy for him to get on the flight and how he might take leave soon instead. However, I didn't hear much because I tuned him out. I thought he was coming up with excuses on why I wouldn't be able to go when suddenly, I heard him.

"You know the stipulations for us letting you go on leave?" he asked.

"Roger, sergeant." I remember looking down at him while he was sitting doing paperwork and him looking up to me, asking me that. I remember looking at him with conviction, but it was with conviction that I finally won the war.

The mental warfare was coming to an end, and I won. He probably took it as conviction in the sense of me coming back to sit in pain behind a radio, but that ship sailed after getting kicked off the sniper team. I told them what they wanted to hear; they gave me no reason to be truthful since they were never truthful.

That night, I was going to fill up the generators that powered the TOC and was getting ready to head out the back door that usually wasn't open since it was closer to the generator's location. Headed towards the opening, I got about ten feet from it when a lieutenant colonel stopped me.

"Hey," he cried. "Don't go out that way. It's not a door; it just happens to be open."

Wow. I thought. *Are you kidding me?*

I turned around and walked out the other way to the generators. Why the fuck did all the officers in the unit act like they had their dick in a vice? Some of the most unbearable worthless fucks on earth were all conglomerated into one place.

On October 22nd, they told me I would be filling out leave paperwork; I wasn't sure if it would finally happen, but it did. On October 24th, I was to fly out, and on the way to the staging point for the flight, I ran into Jack. Our conversation shakes me to my core to this day.

"Perez! Man, what have you been up to? I've been wondering what you've been up to."

"I'm heading for R&R right now!" I replied. "Finally headed out of this bitch, hopefully. How are things with the sniper team?"

"There is no sniper team anymore," he said. "Albie broke us up and put one sniper in each platoon."

"WHAT!?!?" I cried.

I don't remember anything after this part of our conversation, and I believe that I made the same face Richard Sherman made when the Seahawks threw the losing play at the one-yard line in the 2015 Super Bowl. It's a regularly used GIF, so you've likely seen it, but if you don't know the face, it's one of ultimate disbelief. I was so angry inside because had I known that fucking idiot was going to disband the sniper team, I wouldn't have taken it so personally.

I was sitting there, waiting for my flight, but the surrealism wasn't there the way it had been many weeks earlier when Callahan told me that I should fly out for surgery. I fought hard to get to that point and to win the war; I'd have to get my ass on a helicopter headed for KAF. Sure enough, the Chinook landed within the hour, and I was off on my first helicopter ride.

After a short flight, we were back on KAF and escorted to a tent city where transients slept on their way to and from the states. I don't remember how long I was there for; it was anywhere from that night to the next day, but more than likely it was just that night. I remember getting to the tent with my backpack then quickly leaving my stuff while I looked for Tracey, who got blown up with Adam less than a month prior. After an hour, we ran into each other, and I was so happy to see him. We started talking, and I could tell something about him was different right away.

He showed me all the medication he was on, from opiates for the pain to Ambien for sleep. In less than a month, the Army made him a zombie, and that was just the tip of the iceberg for what was in store for him after that fateful day when we lost a good friend.

Before my flight, Tracey gave me an Ambien because he wanted me to get high with him, so I took it. After about forty-five minutes, he asked if I felt anything, and I didn't, so he gave me another, which I immediately took. It was my first time taking Ambien, and I expected it to be like the over-the-counter sleep aids available, but I was dead wrong.

Within the hour, the effect of both Ambien began to kick in, and I was high as a kite. I remember the world becoming hazy, but I was feeling good, and that was all going to come to a screeching halt when it came time for me to get on the bus to get on the flight out of Afghanistan.

I got on the bus, and there weren't any more seats available, so I had to stand. The driver was a soldier they sent from Frontenac, and they made him a driver because they no longer trusted him in any other position.

When he was on Frontenac, his job was to take the personal belongings of the deceased and injured and ensure that everything was secured to send back to them or their families. They caught him stealing their high-value items and sent him to drive buses as his punishment. I can't be sure if that was the only punishment he received, but it isn't out of the realm of possibility, especially with this unit.

He began to drive us to our destination, and he was driving like a dickface, probably out of spite for the situation he put himself in. He was taking hard turns back and forth; within minutes, I became incredibly nauseous, with the entire world spinning.

I was able to hold it together until he parked, and thankfully I was standing in the front because as soon as he opened the doors, I ran out and puked everywhere. It sucked because I wanted to feel good and fall asleep on the flight I would shortly be getting on.

Tracey gave me two other Ambien for the flight, and I was careful with how I ingested those. Other than our backpack, the only other things we were allowed to bring were our helmet and body armor, so I brought only one change of uniform, my laptop, a hard drive, and a few other small things I could jam into the bag. I knew that I wasn't coming back, so having to leave most of my military and a lot of my personal belongings there killed me because I knew that I would likely never see them again.

CHAPTER THIRTY-ONE

THE TWILIGHT ZONE

I had them fly me to Corpus Christi first. When I landed, I went to the naval base to see if I could talk to someone there about my shoulder issue. They told me I'd have to talk to the people dealing with it, so I paid for a one-way flight to Seattle within a few days.

I figured I could try and get everything straightened out while still on leave. When I got to Fort Lewis, I went back to our normal barracks and battalion area because it was a weekend, but most of North Fort was a ghost town when I got there. I walked around the entire area and tried to find out where anyone was. I couldn't imagine why our battalion rear detachment would be anywhere else than the battalion area we usually operated.

With nowhere to sleep, no ride, and no one to contact, I walked around the barracks until I found my way inside. Someone left one of the doors unlocked, so I went in and walked on every floor, trying to find someone. It was empty, so I picked a room and made myself at home.

Since I could not find anyone in my unit, I went to Okubo, the medical clinic, to talk to someone about my shoulder and see if I could find out where the rear detachment was. I saw a civilian doctor, and he ran me through the same gauntlet of strength tests I had been through before and told me I'd probably have to go back overseas because I wasn't hurt.

I looked at him, unsurprised at the cog I knew I was dealing with, and I told him that I wanted to talk to a mental health professional. I sat in the lobby, and after some waiting, they finally called me into the office of the woman who worked there named Elena. Elena was an older woman in her mid-seventies, and she was great at her job. I explained everything to her, literally everything I went through in the last eight months leading up to meeting her, and to end it all; I remember calmly telling her, "If I get sent back to Afghanistan, it's going to be bad for me, and worse for them."

Whatever she did to keep me from going back was a blessing. I started dealing with a civilian doctor, and she especially knew about the bullshit I was going through because she was a retired officer. She was the one I dealt with for the most part from that point on about my shoulder, and she was a great help. I brought up my eye twitching from the stress I was under, and she explained that she experienced the same thing when she was serving, but it went away. She said the twitching in my eye would eventually go away, too. I didn't believe her.

Elena then put me back on a temporary profile and signed me up for a rehabilitation program there at a location near Madigan Army Medical Center. Later that day, I found where our rear detachment was, and it turned out to be in one of the sister battalions areas opposite of where we usually occupied.

I went over to talk to the rear detachment commander and first sergeant, who were captain Gibbler and some staff sergeant I didn't know, who was from a different company. I showed up to tell them what was going on, and they tried to get me right then and there to say that I would go back overseas.

They were so sure that I was going back in that first week that they didn't bother with me too much, but as the end of my leave was approaching and it was clear I wasn't going back, they made my life more and more hellish. They called me into their office almost daily and tried to get me to say that I was going to Afghanistan or something that would incriminate me. They

tried everything from bullying to bargaining, but I wasn't going to listen to any lies they constantly spewed. Because I was on profile, they couldn't make me do anything physically taxing, so all they had was to call me in to their office to engage in mental warfare.

In one of my first meetings, Gibbler tried to get me to say something that would give him ammo.

"Okay Perez," he demanded. "Tell me what happened with your leave. Why didn't you go back to Frontenac?"

"Well," I said. "I flew into Corpus Christi to spend time with my family. My shoulder hurt so bad I went to the naval air station's medical center, but they said they couldn't help me. So I got a one-way ticket here and tried getting help for my injury."

"Where did you go when you got here?" Gibbler asked.

"I initially came to our battalion area, and when I couldn't find anyone, I walked around the outside of the barracks. While walking around, I found a door open and went in to see if I could find anyone, and when I couldn't, I just bedded down for the night in a random room."

"Those idiots left one of the doors unlocked?!" I could tell he was getting heated. He crossed his arms to try to contain himself. "Alright, so then what?"

"I went to Okubo to get my shoulder looked at and wanted to talk to someone about what I was dealing with overseas."

"So you did what they told you not to do before you came on leave, huh?"

"I was seeking aid for my shoulder, that's it."

I wasn't going to mention anything about the mental health visits directly. I knew that if I did, they'd try to find a way to use it against me.

It was rough; many times, I mentally prepared myself before going to war in that office because I didn't want to lose my cool. I forced myself not to do or say something that would give them ammo.

Most of the time, when the bitchfest finished, I'd be so enraged that I would walk two hundred yards to Okubo and wait there until Elena could take me in so I could vent to her. It seemed as if she was the only person who didn't do everything they could to make my life hell—there was a point in which command didn't let me go down to Texas to get my truck. I had to explain that Celina needed a vehicle to look for work, and they said it was too bad, so without their permission, I bought Celina and myself one-way tickets for a Friday evening, and we flew down to Texas.

As soon as we got there, we loaded up in the truck and began the 2500-mile journey we needed to do over the weekend. We didn't get any hotels and instead did some alternated driving. On Monday, we made it just in time to drop Celina off at our new apartment and for me to get to work. I was there just in time for Monday morning's accountability formation, and I couldn't wait for it to end because I was dead beat from that drive. After the morning formation, I went to the barracks and slept until the next formation.

Before we left for Texas, when I finally found the unit, I spent a lot of time in the barracks with the other guys on rear detachment. During one of the nights, I was in the room of some guys I didn't know, and they started to pack a glass pipe with something that looked like weed.

"Is that weed?" I asked.

"No, it's spice," one of the guys replied. "It gets you high like weed but doesn't come out in a drug test."

"No way!"

At first, I thought they were troublemakers trying to get me in trouble, but they reassured me that they had tested since smoking it and hadn't failed. I grilled them some more on where they got it from and to see the package. After confirming that it didn't look or smell like cannabis, I gave it a try. I got high and couldn't believe there was now a legal substitute for cannabis. Though it didn't quite feel the same, it was enough to relieve my stress a bit, and I planned to purchase some of my own. I smoked that nasty shit the entire time I was on rear detachment, and as much as I grew to hate it, I needed it to help keep me sane.

Due to the copious smoking I was doing, I don't remember my conversations with Captain Gibbler. However, I do remember talking with that staff sergeant.

"Sergeant," I said. "I'm not going back because I'm hurt. I don't care about rank; I don't plan on making a career out of the Army the way you have."

"A career?" he responded. "I'm done; this is my first and last enlistment. I have no intention of staying in."

I could tell the unit had worn on him the way it had on most people with a conscience, and we talked man-to-man after that. I also couldn't believe he had made staff sergeant in his first enlistment. If I remember right, he was one of those who signed that infamous six-year contract, but I still couldn't believe that they'd put someone with basically no real leadership experience as the acting first sergeant for rear detachment. That was likely why I primarily dealt with Gibbler because he had no problem making life hell for as many people as possible.

By the time my official leave ended, it was early November. Around this time, I met a guy named Mann. He and I had the same rank of specialist, but he was a medic who worked at Okubo, and they put him in charge of all the medical records for the guys who were hurt and killed overseas. I could tell he was competent—the fact that he was assigned the critical position, which likely should've been a sergeant's job, reinforced that thought. I'll never forget our first conversation.

I described to him what I had gone through over the last year, and because we had just met, I could tell he took my story with a grain of salt. While explaining the situation, his demeanor changed from receptive to disbelief, and he looked at me as if there was no way my story was true.

"I'll tell you what," he said. "Because I'm in charge of everyone's medical stuff, I can look up your file and see what it says."

"Okay great! I'd love to know what it says about my injury." I didn't blame him for not believing me—the whole situation is still hard for me to believe to this day, and I'm the one who went through it all.

After a few days, we ran into each other again, and the look on his face was priceless; it was as if he had seen a ghost or witnessed a serious crime. He genuinely looked shook, and the first thing he told me was,

"Dude, you were right. I looked into your record, and ortho never cleared you to deploy."

He was shocked at the validity of my story, and I was now even more pissed that Cleve had lied the entire time about talking to ortho. We became friends, and I talked to him fairly often when I'd run into him to see how he was managing the workload he had, as many people were dying and getting hurt.

Mann was also good friends with Elena because they both worked at Okubo, and they had a great relationship. One time I was in Elena's office, we were in the middle of a "session," when suddenly Mann walked in with something to tell Elena. He apologized for intruding and said he could come back later, but I quickly stopped him and told him that if it wasn't private, I didn't mind him telling her whatever he needed to.

He walked in, they started talking, and while they were, I was off in my mind, likely replaying whatever led me to her office that day, when suddenly I saw them hug. I watched them embrace; they said they loved each other, and he said goodbye to me and left. I couldn't comprehend what I just saw, not that there was anything even remotely wrong with what I witnessed. My mind couldn't process seeing a soldier in uniform happy to be at work where he made such a strong bond with a civilian coworker!

Seeing Mann happy in uniform at work was a simple scene, yet surreal. This happening with a civilian made it that much harder for me to understand. I realized then that I also made friends with two extraordinary people and was glad they were on my side.

One day, when I ran into Mann, he told me he was going overseas to "help out over there." My heart sank.

I immediately started to plead with him. "Don't, Mann, please don't. You're doing more for the unit here. You're competent at your critical job. You're doing more for us here than you could do over there; please don't go; it's not going to be what you think."

I was doing everything I could to convey the level of fuckery going on overseas, and he wasn't having any of it.

"I need to go overseas," he said simply. "They can use me over there."

While I agreed they could use him over there, I knew they wouldn't use him properly.

I didn't want him to go because I knew it would change him, and I didn't want that to happen to someone with actual values. I understood why he wanted to go: he wanted to help the best way he could and felt that overseas was the way to go, but I didn't want him to fall into a situation as many of us had.

I don't know exactly what happened to him, but he left before the year's end, and I only had Elena to confide in at that point, besides Celina. Besides the relationships I'd built there, rear detachment sucked. It mainly consisted of guys who got blown up overseas, except for Tracey somehow. They sent him to the Warrior Transition Brigade (WTB) on main post. He was there alone and already addicted to pills.

We did everything we could to get Gibbler to bring him back to the unit, but they didn't give a fuck and just left him there to rot.

On rear detachment were other guys who went AWOL before the deployment or didn't pass their physical/mental evaluations. When I first got to rear detachment, only a couple of new soldiers came straight from basic training. One of them wanted out of the Army as soon as he got to our unit, and they were doing his paperwork to get him out, so he just hung out with us and did most of the bitchwork.

Most of us were on profile, so we weren't allowed to do anything, and it was everyone else who did the heavy lifting. We did things like sweep, mop, and other small tasks, but for the most part, we didn't do anything but go to appointments because most of us were also transitioning out of the military.

At some point, Sergeant Major Driver flew into the country, and he was going to stop by to give us a pep talk, so we had to make sure that everyone from the unit was there for his motivational speech. Around thirty or forty of us showed up that morning, most of whom were there because of being blown up overseas. Driver walked in, and the first thing he says is,

"Who here wants to stay in the Army?"

A handful of guys raised their hands.

"Okay," he said. "You guys are free to go."

They got out of there as quickly as they could. Once they left, Driver turned to the rest of us.

"Okay," he boomed. "The rest of you are a bunch of cowards. When the unit gets back, I hope they beat the shit out of you. Tell your wives they better get good jobs because you're not going to have the benefits of the military anymore."

I couldn't believe what I was hearing. Not only did Driver call a bunch of guys injured from overseas cowards, but he also hoped that they would physically assault us, too. I didn't understand what was going through his mind to call guys with purple hearts cowards, but there is no excuse for the senior enlisted person in charge of a battalion to tell that to his men. He continued talking down to us for a good five or ten minutes and then left. He would soon be back overseas riding his ATV around the FOB.

As I sat there, tuning out his speech, I felt as if my mind was melting. This was the man in charge of the battalion, and somehow he concluded that the injured men were cowards. Cowards? How the fuck could he come to that conclusion? It didn't make sense, and the more I thought about it, the angrier I got. I tried to reason with myself for a bit, thinking that maybe he was directing it to those on rear detachment who didn't deploy with the unit, to begin with. But that train of thought was quickly squashed when I thought about the men he excused before the bitch fest.

The rage built the more he bitched, and I was on the verge of exploding. Like always, I bit my lip and just listened intently. Of all the things I have recorded, getting this on audio would've been the best of all recordings. I went wrong using my iPhone to record when I got back to the states rather than the voice recorder I bought and used during and before deployment. Within a year of me discharging, the iPhone took a crap on me after an update, and I lost everything I had on that phone. I was thankful that I pulled the pictures prior, but I don't think there was a way to remove the audio from the app I used on my phone as this was 2010.

The second he released us, I went straight to Okubo. I didn't wait to see if there was a follow-up formation from someone in charge of rear detachment; I didn't care what anyone had to say. All I wanted was to go to Okubo and talk to Elena. It's strange because I can remember the walk over to Okubo, probably because brain matter was leaving a slug trail as I walked there that day. I remember being so fucking angry. I needed to vent, and thankfully I was able to sit with her and chat about what had just happened.

I spent a good amount of time with her and felt better after leaving like I usually did. Elena saved me, not only from going back overseas but also from flipping out on the people I had to deal with on rear detachment. She was so much more than I could've ever hoped for, and without her, I wouldn't be able to articulate my story in the way I have in this book.

CHAPTER THIRTY-TWO

THE VEIL FALLS

Rear detachment was a time to get high and play video games until it was closer to when the unit was supposed to return. During my time on rear detachment, I also tried to stop the combat and hazard pay I was getting for being deployed. I told the leadership to stop it, and they wouldn't.

I then tried to stop it by going to the "soldier readiness processing center," where we did our immunizations and paperwork for deployment. They said the unit had my files, and they would have to be the ones to take care of it. I even talked to a chaplain at the processing center to tell him about the situation; he listened but offered no real help.

I finally thought I got some help when I called IG and talked to someone about the situation, and the lady I spoke to said she would contact the unit and take care of it. When I got my next paycheck and was still receiving combat pay, the following check came, and I was still getting the extra pay. Again and again for a couple of months; this happened before I called IG back.

"I talked to Captain Gibbler," she said. "He said he was going to take care of it."

It turns out he never did, and he, along with the acting first sergeant, went overseas and was replaced by two even more incompetent, empathy-free individuals. I lost all faith in IG at that point and stopped bothering them too.

There was nothing I could do. No one was helping; even the people whose sworn job was to investigate and fix things were doing nothing. I kept collecting the combat pay and did my best to save as much as possible, but all the expenses of my last-minute deployment and return made things more expensive than they should've been.

When the new rear detachment command took over, they made life hell for me for a bit, and you'll never guess who it was. The commander was Captain Cox, and as the first sergeant was Randall! They weren't too hellish, mainly because the deployment was already halfway over. I'd been on temporary profiles for almost a year by this point, so I brought up making it a permanent profile with my doc, which she did. With that came what is known as the Medical Evaluation Board or MEB, where they try to determine if you're capable of doing a different job in the military after getting hurt. If not, you're medically discharged. You can be medically separated with one paycheck and your freedom or medically retired and get paid for the rest of your life.

When I began the process, I was excited because it meant that I was more than likely retiring medically and getting two checks a month for the rest of my life. I started the process by going to some of their mandatory meetings, describing the process and what it entailed.

In the meetings, they explained that they started a new program that worked in conjunction with the VA, which would give the person both their Military and Veterans disability ratings simultaneously. They said the anticipated time of going through it all was roughly two years.

I slumped back in my chair. Two fucking years? There was no fucking way any amount of money in the world was going to be worth sticking around these fucks for the next two years, so I began going all over Fort Lewis to try and find someone to take me in while medically discharging.

Other guys were able to get out of the unit in fear of getting assaulted when the unit got back, and I tried my hand at trying to get out for a different reason. After weeks of looking, I might have found a place because they were looking for a sniper-qualified person to help with marksmanship training, but they didn't want someone on their books who was actively getting out. After that, I realized that no one would take me out of the unit to be medically discharged. I decided to stop the MEB process. It hurt me to know that I was forfeiting a lifetime of benefits. Still, chances were good that I would've been dishonorably discharged or in military prison for saying or doing something to any of those fucks that gave me problems.

The people who dealt with the MEB couldn't believe that I was doing it, but I assured them that if I didn't, they would likely see me in the news over the two-year wait. I waited and waited for them to make my profile permanent, and when they did, I couldn't even follow through because of the shit environment that place was.

From the bottom to the top, everyone wanted out of that place. Only a fraction of them reenlisted, and some stayed in the unit and did the second deployment in 2012. I could've stayed in the unit and MEB without too much drama, but that wasn't a risk I was willing to take.

Having dealt with Randall over the past year, I felt comfortable approaching him when I had a problem, which was the only upside to having him there. I didn't do it often, but one of the things I did ask him was to ask Captain Cox if he could put his signature on a paper to get tuition assistance and start school without using any of my GI Bill.

I asked him, and he forgot to ask for a few weeks, so I kept bringing it up with him, hoping that he would ask for me. I reminded him again one morning.

"Okay Perez," he sighed. let's go upstairs, and I'll ask him."

We went upstairs and waited in the conference room area with offices around it. Once Cox came out of his office, Randall asked him.

"FUCK NO!" He responded loudly. "Perez isn't one of my soldiers."

I was technically still on the books overseas, and that was his reasoning to deny putting his signature on a paper for me. He wanted me to be one of his soldiers overseas, though, because when they first put me in the TOC, he approached me while I was working one evening.

"When you're ready to come back," he said. "You can come back to my company." As if I was playing games like the rest of them liked to do!

I started to heat up when he told me fuck no, and I couldn't wait to get the fuck out of that place. I got tasked out for flag detail, which had my buddy from 3rd platoon days and I going to main post every morning and evening to raise and lower the flag for the day. I had to drive my truck on the detail, which got expensive after a few weeks.

I think Randall was relieved of his command or shared it with someone else. Still, I remember pleading with the leadership to take me off the detail after a few weeks because I couldn't afford to pay the extra couple hundred in gas as I drove a truck and asked them to have someone with a more fuel-efficient vehicle do it.

That's when I got the lecture.

"You remember the big blazer I used to drive?" the new rear detachment first sergeant captain said. "Well, I got rid of it for the same reason, and that's why I drive a motorcycle now. You need to suck it up."

I left that conversation angry. I wasn't doing a fucking road march or a five-mile run, it was a dumbass detail I was on, and they could've easily replaced me with someone who had a better vehicle for the job or someone who even lived on post. I need to suck it up?

I also got counseled by the sergeant in charge of the flag detail because there was a time I didn't show up. However, he took the time to be sympathetic and understand my situation. I got the sense that he was just doing his job as the sergeant in charge of the flag detail.

Eventually, the detail sergeant relegated me to be a road guard because I'd gone to sick call to get a shaving profile. I had enough breaking out from ingrown hairs due to clean shaving, so I went for a profile. That was a new level of hell I had brought upon myself, but it was another small victory for me as it was something they didn't like and had no power to control.

When it was getting closer to July, we had a lot more work thrust upon us, most of which was just "area beautification," janitorial, or landscaping duties. They had guys on profile who had been blown up overseas pulling weeds, and some of the sergeants paid out of pocket for equipment to use to get the job done.

They also liked creating extra work for us by doing things as inefficiently and backward as possible. One example of this happened one evening when Randall decided he was going to use the riding mower to cut the grass around the barracks. Rather than mowing the grass in a way that would throw the trimmings away from the walk area, he mowed it as he pleased and had us follow with push brooms to sweep the fresh-cut grass off the concrete.

I couldn't help but take a picture of the irony

Another time, Cox told me to pour gasoline on the weeds because they didn't want to buy weed killer, and I was seriously considering doing it since he ordered me. I was going to do it in the hopes of getting him in trouble for giving me such a stupid order, but I didn't do it.

One of the guys on rear detachment became our hero in a small way. He got sent back from overseas because they caught him drinking alcohol, which isn't allowed, and in his drunken stupor, he called sergeant major an "Apple-Headed Fuck." We didn't know how or why his punishment was being sent home to the states, but he was someone I often used to point out how idiotic my situation was. There I was with a severe shoulder injury, and they were making my life hell to get me back overseas, yet there were non-disabled guys who they quickly could've sent in my place.

Sergeant Major Driver was one of the first people back from the states, and Randall tasked the overseas ethanol drinker to be the one to go and pick him up from the airport. He laughed, "I don't give a fuck; I'll pop anxiety pills and go get him."

When they got back, I was at the battalion area, trying to avoid Sergeant Major because I didn't want him to bitch about me needing to shave. On the way out of the building, he saw me,

"You're going to need to shave, roger!"

"Sergeant Major, I have a profile…" I responded.

He interrupted me, "You're going to go and shave, roger?"

I stayed quiet.

"Roger?"

I kept my silence, and he yelled, "ROGER?"

"ROGER, SERGEANT MAJOR!"

He turned and walked out of the building. Everyone standing around was taken aback when they saw what happened, and everyone told me he was full of shit because I had a profile, and he couldn't make me do anything in the same way they couldn't make me do anything physical. I didn't shave.

A couple of days later, I was at Evergreen Community College in Steilacoom because I was setting up my future after the Army. As soon as I got there, they called me back to work because "Sergeant Major wanted to talk with me."

I knew that fucker didn't have anything important to say, but I had to go back, so I did. I pulled up to the battalion area, and he was there outside the battalion waiting. When he approached me, I stood with my hands behind my back at parade rest.

"So you're Perez," he said. "I've been told that you're not a stellar performer."

I stood there silently as he continued demeaning, which was only a fraction of the time compared to when he called everyone cowards a few months ago. I'd gotten so used to this same speech that everything he said after that went in one ear and out the other.

CHAPTER THIRTY-THREE

NOT A STELLAR PERFORMER

They had us cleaning the battalion area, still preparing it for the arrival of everyone. The only people who used the battalion building while everyone was gone were the families—mostly the wives—of a lot of the leadership. They used it as a place for their "family readiness group" to meet, but instead of cleaning up after themselves, they left that restroom nasty for most of the year.

The place smelled horrific; it may have had plumbing issues though everything flushed. No one could muster themselves to clean that place out, so I took the initiative. There was a gas mask left behind in the battalion area, and I told the sergeant in charge that I would use that gas mask to clean it, and as a reward, I was going to take possession of a picture that Morton had on the battalion building wall. It was a picture of Alpha Company near our Strykers taken at YTC during one of our times out there. A regular 5x7 slipped into a plastic protective spot on the wall with other photos from the years before deployment.

No one gave a fuck about anything, so I put the mask on and went in with as many heavy-duty cleaners as I could. I hit that place wall-to-wall with everything from tile scrub to bleach, and after I finished cleaning, it still carried the aroma of a rotting asshole but was a bit more tolerable.

When the unit returned, they had to undergo a post-deployment process of paperwork and mental health evaluations. I went there to do paperwork, and one of the papers they gave me asked me when I got back to the states, and it had to do with combat pay.

I sat there after reading over the page; I couldn't believe the Army was asking each individual when they got back to the states so they could know when they should've stopped combat pay. I could've lied on the paperwork and put down the date everyone got back, and likely no one would've known any different, and I would have received a year of combat pay when I was only there for a few months since they didn't stop it when I came back.

I contemplated long and hard about what to put down. I wanted to lie, but I also didn't want to prolong my date of getting out that was quickly approaching. Instead of doing the wrong thing and lying, I put the correct date down, even though I knew I wouldn't get paid over the next few months. It sucked because I wanted to lie so badly. Still, my situation with not getting paid was better than getting caught lying and possibly getting something other than an honorable discharge.

Now that my platoon was back, we held a party at one of the sergeant's houses. With no more paychecks coming in and no way to pay rent, I didn't know what to do, so I decided to approach my friend Han that night. I asked him if Celina and I could live with him for a bit because we were about to be homeless. I hated having to do it, mainly because he had issues of his own, but there weren't too many other people I could've asked for help. He said yes, and we moved in with him shortly after.

They got back around mid-July 2010, and I went on terminal leave in August 2010 because I had enough vacation days saved up that allowed me to get out two months earlier than my actual end-of-contract date in the Fall.

I'll never forget signing out for the final time. I remember the sergeant on staff duty at the battalion headquarters where I signed out. I don't recall our short exchange as I signed the paperwork, but I'll never forget driving off and looking at those buildings and the unit area one last time. I knew it would be challenging, mainly because they left me in a hole, but my life was mine again. Finally, I was free to chart my path the way I saw fit, and I was eager to start my academic journey.

Glad I took this picture to share with you all. I'm the soldier on the very far right!

I was thankful because I started school in August at Tacoma Community College. Still, it sucked because I was supposed to collect money for housing through my GI Bill, but I didn't get anything because I was still technically on active duty.

I went the next few months with little to no money until October came, and I was officially a civilian. When I started getting paid again, Celina and I moved out of Han's hair and rented a house in Tacoma. Celina was working, and I was going to school. Things were looking like they were going to get better.

Living in that house was a pain in the ass; the mattress almost didn't fit upstairs, and I had to duck to walk up! The renovated attic was the master bedroom, and the living room of the house was 70 percent single pane glass by surface area, so it was freezing during the winter. However, it was a home I knew I could always come back to and relax without any bullshit.

The first semester of school was quite the experience since I was still traumatized by my military service. When some of the students would get rowdy during the lecture, I would anticipate yelling from the professors that, of course, never came. I did reasonably well even though I didn't apply myself 100 percent. Most of what I wanted to do was get high, play games, and live stress-free. I didn't want to get into academic probation either, so I did my school work and treated it as a job, in a sense.

That New Year's Eve, we spent in downtown Seattle with Duncan, his wife, and a friend of theirs. We tried to spend it together, even on active duty and went to the Pink Floyd laser light show at the science center right next to Space Needle, where they'd pop fireworks at midnight. New Year's Eve laser light shows are scheduled to end right before the fireworks, so you can immediately come out to enjoy a close view of the firework show.

I remember around that time that Celina and I talked about having kids. We were having this discussion for a while, but at this point, she wanted to start trying because she said it was going to take time to get her pregnant. Well, guess what? She was wrong. I knocked her up quickly, and we took a fantastic journey to parenthood over the next nine months.

January 2011 marked another historic day: the end of our lease to that tiny crappy house in a shady neighborhood. After that, we moved onto an acre lot in a decent spot right next to a grocery store and a hardware store. It was nice because we often walked to the grocery store to get our food so Celina could get her exercise in to keep the baby happy. We also attended birthing classes and did everything to get the house ready for the baby.

One night in March, the craziest thing happened. We were dead asleep; it was around 4 a.m. when suddenly we were awakened by every dog in ears distance barking. It sounded as if every animal outside was making noise. It was surreal; I was half asleep and couldn't come up with a reason for what was happening, and after a few minutes, there didn't seem to be anything urgent, so I went back to bed.

The next morning, every channel was reporting a powerful earthquake that struck across the Pacific Ocean off the coast of Japan. Then everything made sense. The animals' sensitivity to their environment made them go on the fritz when they felt the jolt that happened on the other side of the ocean.

The following days were intense, with attempts at saving the cores at the nuclear power facility from overheating and melting down in Japan. I couldn't help but want to go and assist in the cleanup and recovery efforts since at the time, I was toying with the idea of starting a nonprofit organization that dealt with disaster relief in an attempt to work for myself when I got out.

I felt awful about the disaster, though—there were a lot of us, including myself, who had good training, medical and otherwise. I also knew that many of us enjoyed what we did in the army, but we just hated that the people in charge had trouble with basic tasks. Many of us, myself included, likely would've re-enlisted had we served in almost any other unit.

But there wasn't much we could do in terms of helping the people of Japan with the disaster they had just gone through. Like most disasters since then, all we can do is stand by and watch like most of the world. So with no financial support, I stayed put.

CHAPTER THIRTY-FOUR

FROM SPARK TO FIRE

My first daughter Adelynn was born in November 2011, and the series of events after her birth is something I'll never forget.

I finished my third and final semester at Tacoma Community College in December. It was an incredible place for me to get back into an academic setting. They continue to do great things for their students and community to this day.

Since I had completed three semesters there, Celina wanted to move back to Texas before Christmas; that way, the baby could spend her first one with the family. We broke our lease, packed my truck to the brim and parked it at P's house, and then jammed my 99 Saturn SL1 to the brim with stuff. The car was so packed that I put the driver's seat completely forward to make as much room for crap as possible and had to get into it like a racecar driver: right leg in, shift the body around, then drag the left leg in with my knees basically to my chest.

The trip would be entirely driven by me so that Celina could tend to the baby, and we headed down to Texas. The trip went smoothly overall, and we took five days to make the 2500 miles. When we got down, the aura of the place made it feel like a weight was placed on my shoulders initially. South Texas and I were never made for each other, so returning took some adjustment.

When we moved back, I intended to work in the oilfield because people were making a lot of money at the time. Of course, a lot of hours put in, but I was looking forward to finally getting paid a wage where my family could live a bit more comfortably. I hadn't stopped smoking until I moved down, so I had to wait a while to apply and hope that they weren't going to do a hair follicle test. Most of them did, so options were limited. I applied at quite a few places and was even trying to pick up ranch work because I just wanted to do something I could enjoy, even if I weren't making oilfield money.

When I couldn't find much luck, I went through a temp agency and was able to pick up work at a drug and alcohol rehabilitation center. It paid minimum wage and was easy work because I just had to do things to help the facility's clients. I continued working there while getting clean for a hair test and learned a lot about addicts, including myself.

The most powerful part of the whole experience was attending their AA meetings. There were many powerful stories of how alcohol destroyed people's lives, both directly and indirectly. It made me reflect on how much alcohol played a role in destroying my family and life and how I dug that pit of despair deeper during my military days. The saddest part of that whole ordeal was seeing people who made significant progress go right back into a terrible environment outside the facility walls and fall; so many of them went right back into the lifestyle that got them there in the first place.

I continued applying for oilfield work and was finally offered a great opportunity through a different staffing company. I was going to be sent to driver's training in Oklahoma for a significant oilfield company to drive one of their rigs and was looking forward to it when talking about it on the phone with the woman to interview me. When I showed up to do the interview, everything went well until she said something that floored me.

"We love ex-military," she said. "They're great because we know that you're used to being away from your family, and that's a lot of the life for most of these oilfield men."

Well, I thought to myself. *I just had a baby and left that life for a reason.*

Everything went well. I finished the interview, and I headed home.

On the fifty-minute drive back, I thought about being gone all the time again and decided that going back to school and getting paid to do so was probably going to be better for my family and myself in the long run. When I got to Celina's parents' house, where we stayed when we moved down, I talked with her, and we agreed that I would go back to school. We would then figure things out from there.

I have wanted to learn about plants and their incredible versatility since as early as I can remember. We started looking at the programs offered at Texas A&M University-Kingsville (TAMUK) and were disappointed to see they didn't offer a botany program. I don't think that interest will ever go away. Unfortunately, TAMUK had animal science and outdoor types of degree plans offered through their outstanding wildlife department, but those didn't seem like a good fit. I figured that the next best thing to botany would be chemistry since it often deals with how plants interact with their environment.

I did what I had to do to get back into school using my GI Bill. Then, in the fall semester of 2012, I was back in school after an eight-month hiatus. Unfortunately, the excellent skills I was building in algebra had started to slip away during that time.

When I was at TCC, I did the first of two classes that they considered precalculus. When I got to TAMUK, the advisor looked over my transcript.

"Well," he said. "You took pre-calculus. That's good—we can put you in calculus."

I died a little when the advisor said that, but I wasn't going to say anything because I didn't want to use any more of my GI Bill than I already had to. I was shocked, but what he said next after a couple of minutes is what hurt.

"Okay, there were twenty credit hours you took that didn't transfer."

"What, what?!"

"Yeah sorry, the description of the classes doesn't match up with what we have in our catalog; those will be the ones that won't transfer."

Calculus I and General Chemistry II were the most challenging classes I took in my first semester back, and it didn't turn out to be as bad as I was expecting them to be. Calculus and General Chemistry II were challenging, but there was an extra hurdle I had to overcome at the beginning of the semester for my chemistry class.

My professor, Emi, was from Germany, so her accent was heavy, and it was my first exposure to a native German speaker. Trying to understand what she was saying while she was teaching made the class a bit more complicated.

In the first few weeks, I would leave her class with my brain physically hurting from concentrating so hard on the course content while also trying to figure out what she was saying. As I got used to her accent, the slight headaches I got stopped, and then it was just working hard to pass the class.

I'm glad that I had Emi during my first semester because she gave me a reality check as to what one can expect from university life and the pursuit of being and growing into a scientist. It was a tough class, and a classmate and I would go to Starbucks after every class and study the material we'd just gone over. We worked our asses off and ended up with As in the class, so it was worth all the pain and suffering.

In the spring of 2013, I took Organic Chemistry 1 with Emi, and this time she asked if anyone wanted to volunteer in her research lab.

Wow! I thought. *What a great opportunity!*

I didn't know what she researched or what she did, but all I knew was that I wanted to get into a research lab to get experience. I thought they had to offer these positions, not advertise them as something that inexperienced people could volunteer for, so I stayed after class to let her know I was interested. That's how I fell into chemistry research until the day I left TAUMK.

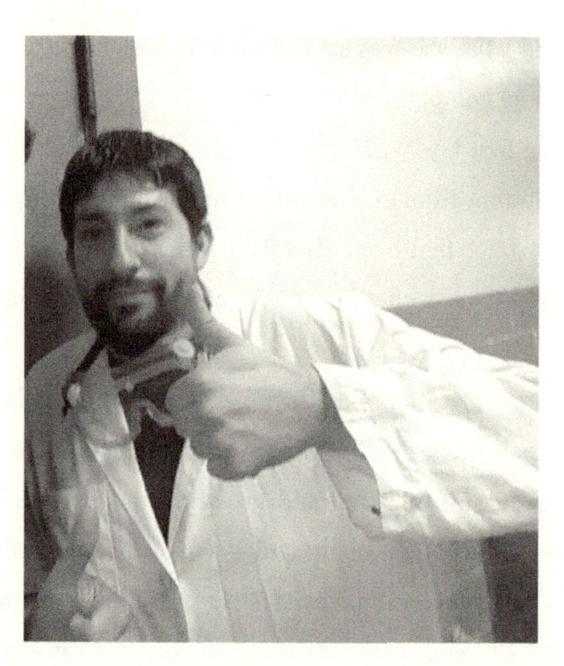

I spent so much time in labs but loved every minute of it.

I would drop my chemistry major that year to do chemical engineering, but when I talked to her about it, she said she would help me if I did the double major, so I decided to stick with both. When I decided to double-major, I didn't know if I could handle engineering classes but knew I would work hard and do everything I could to pass and be successful. I learned a lot about being on a research team and doing research and wish I could've dedicated more time there, but I had many things going on.

This was around when I realized the flame that has burned within me from the start of my life: it only gets stronger with each passing day, and though I have my down days as we all do, I strive always to keep that fire raging.

That semester is also when I enlisted in the Army Reserve and served for another year. I did it for the paycheck but continued smoking weed with reckless abandon because it was the only way I would be able to handle everything I was loading upon myself on top of the chronic shoulder pain that still plagued me.

This time in my life wasn't necessarily a bad experience, though it would've been, had I stopped smoking weed. The chronic pain from my shoulder injury sucked, so I did what I could when they expected us to do PT, but being a broke dick didn't make things easy. Academics were now my focus, and once I got into my junior year, I was discharged from the Army Reserve honorably (thankfully, because the drug test bug almost bit me).

In summer 2013, I spent a lot of time in Corpus because I took summer classes at Del Mar College and started a work-study position through the VA at the beginning of the year, which was just a part-time job that paid minimum wage. I began by helping the veteran representative at the local work source, but he retired not long into me doing it. He recommended that I find a place that services veterans to see if they had other work-study positions.

At this point in my life, I was deeply missing the brotherhood I had built over my years in military service, but the small team working on research helped fill the void that was missing. I was starting to develop a new academic community, which made me grateful for choosing chemistry as a major.

I went to the local county veteran service office and inquired about the work-study spots. The representative said there weren't any, so I asked if we could set some up there, and he agreed, so we worked together to set them up.

I was able to keep my part-time job to continue helping local veterans. I did everything to supplement my income, but things still got rough. Celina had also started school, and I started getting money for housing, so we were able to move out of Celina's parents' house and rent one.

We moved in after the summer, and that was a blessing during fall and winter. The house without A.C., though, turned into a curse once the weather warmed up. As we moved from house to house, renting year after year, our rent got gradually more expensive.

By the fall semester in 2013, I thought I hit the top of the mountain on my climb towards my degrees by taking differential equations, three engineering classes, and two higher level chemistry classes. I was trying to maximize my GI Bill benefit, since it gets used based on days in class, not credits taken.

Term: Fall 2013 TAMUK

College: College of Engineering
Major: Chemical Engineering
Academic Standing: Good Standing

Subject	Course	Level	Title	Grade	Credit Hours	Quality Points
CHEM	3125	UG	Organic Chemistry Lab II	B	1.000	3.00
CHEM	3325	UG	Organic Chemistry II	B	3.000	9.00
CHEM	3331	UG	Physical Chemistry I	B	3.000	9.00
CHEN	2371	UG	Conservation Principles	B	3.000	9.00
CHEN	3347	UG	Chem Eng Thermodynamics I	D	3.000	0.00
MATH	3320	UG	Differential Equations	B	3.000	9.00
MEEN	2355	UG	Statics and Dynamics	B	3.000	9.00

I thought this was the top of my mountain, but it wasn't.

CHAPTER THIRTY-FIVE

A NEW MISSION

My junior year started in the spring semester of 2014. Around this time, things were piling up: so much conversation was around the war with ISIS. In our personal life, rent became too expensive. On top of that, another baby was on the way, so there was no way we were going to be able to afford anything with us both being full-time college students.

I took another crazy class load with five engineering classes and dropped one due to bombing the first test. At this point, I wasn't interested in working the oilfield anymore but was more interested in using my sniper training and getting paid by a private military company. After seeing they sought someone with my qualifications, I applied for a "designated defensive marksman" job. Still, I didn't think I would hear back from them because I thought they were more interested in guys with special operations training.

Within a few days, I got emails and calls telling me they were interested, and I couldn't believe it. I talked about it with Celina, and she didn't want me to go. I didn't want to either, but I didn't want to work in the oilfield.

This wasn't the only stress that came about from being broke. That summer, I went to Tennessee with my dad to visit my aunt, who I hadn't seen in many years. The car she drove for the last decade was finally going through its final days, and her son gave her another vehicle. She was looking to donate it, and I asked her to donate it to me because the little bit of money that I could get from it could help me get by.

On the road trip up, my dad and I talked about my job situation, and he said he would help me with whatever he could until I could finally get my VA disability pay straightened out. At this point, Dad had brought his life back mostly under control. He was saving money and working towards building himself a spot no one could take out at the ranch. The 2008 recession hit Ruperto hard with all the overhead he had, and he ended up losing everything, and thus their house of blow came falling. Since dad lived in Ruperto's old place, they came back and kicked him out. It took him almost a decade, but Dad eventually built himself a nice spot out on the ranch. He also worked to be more of a father than he had before. Thankfully, there were no strings attached to the money, and he didn't even expect me to pay him back, which was nice. It was my old man that kept me from being a school dropout the second time around, and he probably gave me about five hundred dollars a month for six months before I started to get my disability pay.

That summer, my youngest daughter, Amelia Elena was born. Her middle name was in homage to Elena, who saved my life while in 5th Brigade. She mixed things up in our family by being such a smart baby! Not long after she was born, she became aware of her arms even though she didn't know what to do with them. She would force them straight out behind her—almost parallel with her body—and when we would try to get her to put them down with a bit of force, she would resist! When she was around eighteen months, I came home from school one day thinking she knew just a few beginning letters of the alphabet. When I walked through the door, Celina told me, "Babe, you'll never guess what. Millie knows her whole alphabet!" I couldn't believe it, so Celina drew every letter out in order, and though she couldn't pronounce the letters correctly, she could make the sounds for all of them. I was blown away; I couldn't believe that she knew them all!

In the fall 2014 semester, things got a bit easier because my disability pay increased slightly around this time. The VA gave the first percent I was awarded without any evaluation. The appeal process took over a year, so there was a long waiting period, which made things hard because if they had evaluated me the first time, so much hardship could've been avoided.

In winter 2014, Celina graduated with her degrees in Political Science and English. A few months later, she got a job with the Social Security Administration. After she started working there, things got more manageable for us financially, and she did a lot to streamline some of their processes and got great work experience. It worked out because I was about to run out of my GI Bill benefits and have to use my state benefits, which still paid my tuition but didn't have a monthly housing stipend.

When I ran out of GI Bill benefits, I took twelve hours per semester instead of the sixteen to eighteen I usually took, so classes got easier too. Then, spring 2015 was special because we started working on our senior engineering project. However, we mostly spent it getting familiar with a computer program and doing a lot of writing projects because part of our senior project was a writing-intensive course we were supposed to complete before graduation.

This was around the time I came up with a great idea for something I wanted to patent, which dealt with concepts I learned as an engineering student. I'm no expert, of course, but after years of banging my head against the wall, the haze of how everything tied in together started to clear. I began to feel like I was seeing the world in a different way—doing engineering calculations didn't seem like the herculean task it once did.

In the fall of 2015, I stopped by the university's veteran office to take care of the usual paperwork. That was when they presented me with an opportunity. They knew me pretty well because there was always an issue getting me certified due to being a double major. Recent troubles had to do with the cancellation of chemistry classes that I needed due to under-enrollment, so substitutions had to be done, and I ended up taking a Master's level course.

"Hey," they said. "You're an engineering major, right?"

"Yeah! Why what's up?"

"There is a scholarship available to a veteran environmental engineering major, but it's meant for a graduate student, and the guy who originally received the award dropped this semester. If you're interested, go talk to this professor and see what he says."

I was immediately interested. I took the professor's information down and stopped by his office to talk about it. He asked for my resume and research experience, and I told him I didn't have any in the engineering department but that I'd been doing it in the chemistry department over the last few years.

After submitting my resume to him, he went forward with me, and I was awarded the $5,000 scholarship.

I had a year of research ahead of me. I enjoyed doing the research and wish I could've spent more time there too, and I was grateful for the extra money. I was conflicted because I didn't want to spend more time in one research lab than the other, and I did spend more time dealing with the engineering research because it was newer, and I was less familiar with what I needed to do. Thankfully Emi, my German professor, was the director of the lab. She was fantastic and understood the situation, and I did spend a lot of time there as well, but I was getting stretched pretty thin, especially with my senior project underway.

The struggle was real. I was doing research in two labs, 12 hours of senior-level STEM courses, senior engineering project, putting on handgun classes and attending gun shows to seek students, raising two little girls, and taking care of home duties. And though I wasn't doing the GI Bill work-study program anymore, I was still showing up at the veteran service office to volunteer and help with handgun licensing paperwork for older students who had trouble navigating the internet. Though I wasn't taking 16-19 hours a semester anymore, I had just as much on my plate.

That last semester was quite the roller coaster ride! Time continued to feel both slow and fast simultaneously, but finally, spring 2016 was my last semester, and I made the Dean's List. I was happy because I knew I could do that at a minimum every semester, but life for me has never been easy. Hence, it was nice to get at least one gratifying semester in my entire university experience.

We started the semester expecting to get the Pell Grant one last time, but they denied me for having too many hours when I applied.

I appealed and lost, but the document didn't make any sense. I talked to my sister about it because she had worked in financial aid for over a decade. She explained that I was still eligible because I was not on academic probation and waived hours for double majors. I was so glad I talked to her about it.

After that, I met with the interim financial aid director to try and explain my situation and that I am eligible. All she would say was something like, "Just go and Google it and you'll see, look it up."

Realizing that she was never going to listen to me, I decided to call the president's office to try and meet or talk with him. I thought it would be easy because he was an Air Force veteran, so if I called and said that a veteran was having problems at the school, he would reach out and listen to me.

I spent that entire semester calling his office multiple times, being given the runaround, or just blown off. One time I lost my cool and started yelling at the secretary, but I apologized and told her that I didn't mean to yell at her and that she was just a bystander.

After yelling at the secretary, I thought that word would get to the president that a crazy veteran kept calling, trying to meet with him. Well, they probably told him, but he still chose not to meet with me, so I kept calling until the semester got too busy. At that point, I was out of free time.

Towards the end of the semester, though, I started to call back again, and finally, they set up a meeting for me to meet with one of the university's vice presidents.

Finally, I thought. *Someone will listen to me, and we will get this fixed.*

I sat down within the vice president's office a few days later. She asked me to tell her what was going on, so I did. I remember sitting across from her, shifting uncomfortably.

"Okay," she said after I relayed the entire situation. "I know someone who is an expert at this stuff, so if anyone knows what's supposed to be going on, it'll be him."

She picked up the phone, made the call, and put it on speakerphone when suddenly he answered. She spoke aloud, explaining to him the situation.

"Yup," the guy on the phone said. "You're absolutely right. Because you're a double major, you can get a waiver for the hour issue and get your financial aid."

The vice president thanked him for quickly clearing the issue up, and when she hung up, she apologized for the things I had to do to get someone to listen to me finally. She told me she would instruct the interim financial aid director to change the university policy immediately. They would rewrite it so the language more clearly explained that double majors in good academic standing could get waived for their amount of hours, which would avoid further confusion.

Of all the things I did in my time at TAMUK, this was the thing I'm most proud of accomplishing. Getting things cleared up for lower-income students who were double majors was suddenly an issue on my radar, and I could make a difference in someone's life because of my situation. Who knows how many students who were double majors were denied financial aid later in their academic career due to the inability of people to do their job correctly and "look it up."

A few days later, I even walked into her office at the exact moment as she was typing up the new policy. She even confirmed it to me: “I’m working on the new policy right now,” she said. That brought joy to my heart, witnessing the change I had made. It also brought me back to Frontenac and the moment I won the war with those in command. I was there standing firm, looking down at those who tried to hurt me and others on a grand scale, victorious after scrapping for so long.

CHAPTER THIRTY-SIX

EMPTY THE CANS

My shoulder pain went from being tolerable to incredibly painful throughout that semester. Many days out of the week, I'd have to sit and breathe through the pain to be able to endure it. I knew it wasn't right, and since we had great federal benefits through Celina's work, I decided to look up doctors around Texas.

After a few days of researching, I found a good doctor in San Antonio and called his office to see if they took our insurance. Thankfully they did, and within a few weeks, I was in his office. I had sent him the MRI I had done through the VA a couple of years prior, and he had said the same thing the other doctors said:

"The MRI doesn't show anything."

I was instantly disappointed to hear that because I thought I would have to continue in pain. He had me stand up and get ready to go through the resistance tests they do whenever a patient has shoulder problems.

I stuck my arms out, he placed pressure on them, and I was able to resist. I had pain, as always, but could still keep him from pushing my arms down. I stuck them out to my sides, and again he pushed, and I resisted.

I thought the exam was over when he suddenly said, "Okay, now stick your arms straight out and turn your fist towards the ground like you're emptying a can."

I did what he said and didn't think that a slight change in my wrist orientation would make a difference, then he pushed down. Not only did I have incredibly sharp pain, but I was also unable to stop him from pushing my arm down. I freaked out!

"Can we try that again?" I asked. He, of course, obliged.

This time though, I got mentally prepared. I took a deep breath, shook my arms out, and held them out in front of me, prepared to give it everything I had in my body and soul to resist the pressure he was going to apply.

But again, there was nothing I could do to stop him. The pain was too much; my oldest daughter could have pressed my arm down in my condition.

Saying my mind was blown wouldn't be enough to describe how that made me feel. All the hell I endured, being called a malingerer, mistreatment, misdiagnosis, fucking everything. EVERYTHING! I couldn't believe that seven years of Army and VA doctors could not diagnose my shoulder injury all because they didn't tell me to "empty the can."

The doctor said it was apparent I was injured, even though the MRI didn't show anything. We scheduled a date for surgery just a few weeks later. I finally got the surgery on April 13th, 2016, the day after Kobe Bryant's final game. I stayed up late to watch it and ate at a popular seafood restaurant earlier in the day. Celina and I rented a room in San Antonio for surgery early the next day.

The following day for the surgery, everything went pretty quickly. I showed up, got in a gown, and sat in a hospital bed for a while before the doctor and anesthesiologist showed up. The doctor sat next to my bed and told me what was going down. He explained that I would be injected with stem cells as part of my treatment. I was a good candidate for them since my injury was seven years old.

I said that was awesome. A nurse then hooked me up to an IV to loosen me up. I wasn't too nervous; the way I see it is some kids go through worse all the time; I can nut up and take the knife like a man.

The nurse wheeled me to the operating room, and I moved onto the operating table. I remember looking up into the light that got used during surgery, but it was still turned off. With the IV drug kicking in, I was in a daze with the translucent appearance and pattern of the lights. That was the last thing I remember, and the next thing I knew, I was awake, lying upright in the bed.

I looked around, and when I looked right, all I saw was a shoulder all bandaged up; no one was around.

"Wow," I said. "It's done?"

The whole situation felt like I took a thirty-minute nap, not a four-hour shoulder surgery. Not long after, a nurse came in, and soon after, so did Celina. After a while, they brought a shoulder sling, and we were off from the hospital.

I was out of commission for a few days, but still had one month of school left and had a few loose ends to tie up. I continued going to engineering research as much as I could and finished the poster that I had to make for a presentation at the university. I presented it in a shoulder sling, and it was a fun experience. My dad even came out, which was nice.

I was back at the grind within days of my surgery.

I also had to do a presentation in the chemistry department for their first annual department exposé. That didn't go as well as the poster presentation. I hadn't eaten that morning and wasn't so confident in what we were presenting, simply because of all the academic burnout I had at the time. I know I could've done better, but given the grind, I was proud to graduate with my 3.0 GPA as much as another would've been with their 4.0.

The real highlight of that day was seeing Sly again. I called him that semester to let him know the date and, of course, sent him an invitation. I didn't talk to him until dinner, and even then, we didn't talk too much because there were a decent number of families there.

That day also was the most recent time that my mom, dad, sister, and I will be under the same building, which is a crazy thought. The amount of baggage we still carried wasn't discussed, but it was apparent some of it still burned when we were around each other. Since then, there hasn't been a time worth getting the four of us together, and that's probably for the best.

I asked Elena to come too, but she could not, so she just gave me a generous gift in replacement for her trip cost. Sly and I did get to talk, though, and I told him that I would be patenting this great idea I had so we could go on a badass vacation. I was planning on using the large backpay I was due from the VA any month to file the patent, so I was looking at things being good sooner rather than later. That night while we were there, Sly gave me a gift only he could give me. He framed the Green Sprangletop training card we used in Plant ID and gave it to me with a scorecard with our scores from back in the day tucked inside. It was such an incredible gift! I was floored, and we embraced.

"You know the guy who beat you for first place the year after we won state?" He asked me. "Well, he got a full-ride scholarship to MIT."

"Woah!" I exclaimed. "Really? That's awesome!"

He didn't mean the kid received the scholarship from the competition; he just meant that the kid was smart enough to earn the ride.

That was when I just sat there for a bit and thought about how much further I could have gone if I just had a half-stable life. I would imagine that the kid who won came from a happy home, with parents who could afford to pay for the best of the best, which allowed him to pursue his interests and prosper. I certainly didn't have much of a network, which likely contributed to my lower luck.

I don't have any regret about what I've been through. What hurts about it all was that there was never any real equilibrium for me to get used to. Being poor sucked, but it wasn't really what drove my stress and anxiety—it was living with an addict who thought everyone had it out for him.

As a kid, my mother made homemade beans one day and missed one of the small stones that sometimes come in the bag. My dad was the unfortunate one to get it in his beans at the dinner table, and when he bit into it, he began a furious rage. "YOU FUCKING BITCH!" he said over and over. "YOU DID THAT SHIT ON PURPOSE!"

As he got angrier, the memory fades. It was hard to live with someone who thinks the entire world is out to get him. As if living paycheck-to-paycheck wasn't hard enough, I lived with a nuclear bomb ready to go off at a moment's notice.

It wasn't easy to get to this point, but I knew life would get better now that I graduated. I wasn't going to sit there long and dwell on it too much, so I went back to talking to everyone and did my best to spend a little time with everyone.

Celina and I had been applying for jobs back in Washington State; we couldn't wait to leave the extreme heat of South Texas. We had to rush out because the school year was getting ready to start, and they wanted her there a few days before the first day back. She ended up getting a job offer from a school in the Seattle area, so we quickly prepared to move up.

We packed the van up and prepared to set off on the long trip to Washington State. I drove the entire way again because the electric seat was out, but I didn't mind driving on road trips. It's nice to take the occasional break, of course, but I enjoy being the one behind the wheel.

The morning after we stopped for the first night, I got an oil change done while eating breakfast. The guys said I needed new tie rod ends, and they wouldn't recommend making such a long trip without doing it. It wasn't in the budget, so we went off and hoped for the best.

Everything was going smoothly. The girls were doing well, and everyone was enjoying the sights. It wasn't until we got a few miles into Colorado that we had some trouble a few days after leaving. We had just stopped to grab a pizza, and everyone was grubbing out while enjoying the mountain range in the distance.

Suddenly, a loud, high-pitched squeal came from the engine, and I lost power steering. I hit the hazards and started to slow down as I pulled onto the shoulder. Strangely, the noise stopped, and control of the systems came back to me.

I turned the hazards off, and less than a minute after it stopped, it started up again and sounded worse. This time I lost power steering, as well as my mind.

"FUCKING SHIT!" I started to yell loud expletives, and in my fit of rage, I scared my daughter so badly that she threw up and was shaking with fear.

Immediately I felt so horrible and calmed myself to pull the vehicle over. We got her out to clean her off, and after getting her changed, we made a call to get help with a tow. I had Celina walk with the girls while I grabbed a few things when suddenly a stranger pulled over to give us a ride.

He was a Mexican guy who spoke no English, and he had cans of gas in his vehicle, but I was thankful because it was a good ten-minute ride to town. There wasn't much said between us on the ride, but Celina, who speaks much better Spanish than me, did the small talk that took place. Before we knew it, he dropped us off at an auto part store, and it was a convenient spot because it was right next door to a hotel.

I had the vehicle towed to the store, and that night I put it in gear and crept it over to the hotel. After inspecting what was wrong when it showed up, it turned out to be an idler pulley issue that caused the belt to get off track.

The pulley needed replacing, but it was Labor Day weekend, so there wasn't a place open until Tuesday, and this happened on a Thursday. I wasn't going to spend an entire weekend doing nothing but draining extra money on hotels, so I started looking at videos on-line on what I had to do to fix it myself. The whole process seemed simple enough, and with the tools that the auto part store lends out, I was able to do the job in about three hours, so the incident only cost us an additional day.

That was it, and we were off again and finally made it to Washington. It was a pain in the ass for the first few days when we got there, though, because we stayed in a hotel room for around three days and were running out of money. It wasn't until we were just about out of money that I reached out to my friend Jack again out of desperation; I was hoping he would invite us in because he knew what was up, and he did. We were all so glad to be out of that hotel room.

We stayed with Jack for roughly two weeks or so before we got our apartment, and in the time I stayed with him, I learned a lot about the unit that I didn't know.

One of the first things I remember him asking me is, "Hey Perez, did you ever get your medal?"

I didn't overthink it and just said, "No."

We didn't talk too much about it then and there, but I also learned that he never smoked weed until recently.

"Shit!" I said. "If I had known that you didn't smoke weed, I wouldn't have talked to you about it in Afghanistan." Jack could have reported me at any time and may have gotten me into even more trouble than I had.

There were many things I didn't know about Jack and the area, so hanging out with him for those few weeks was great. Once we got our apartment, I was glad to have a place of our own but was depressed about being in an apartment because we were used to having a yard and more space.

I never wanted to stay in an apartment since living in one after returning from deployment. It had been a decade since I lived in an apartment, and this time we were cursed to be put on the third floor.

Shortly after getting into the apartment, HR at the school district let Celina know that her annual salary would be about $8,000 less than what the school told her over the phone before we moved up. She would be making as much as she did at Social Security before we moved, and we moved from one of the cheapest places in the country to live to one of the most expensive.

She tried to get them to give her the roughly $45,000 salary they promised, but they didn't budge, so she put in her two-week notice. That same day, we started applying for jobs. We were worried about being stuck without money for rent now that she was jobless, but not long after she left that job, the VA backpay finally hit our account. It was $40,000, and we had nothing to worry about for the meantime now that we had a nice cushion to float us until we found a job.

Having four degrees in the household, we thought at least one of us would find work right away, but we couldn't have been more wrong. We paid off debts, and the day I woke up and saw that money in the bank, I called to pay our minivan off as well. I also bought about $10,000 of stock in three major companies.

The $40,000 quickly wasn't around anymore, and I still had the expense of flying down to Corpus Christi to rent a box truck. We needed to load up the stuff that was in storage and drive it up to my new home.

Within a week in October 2016, I flew down alone back to South Texas. I did quite a few things while down there, including picking up the newest member of our family in Houston, a Blue Merle mini Australian Shepherd. So after a few days of doing things, I packed up the U-Haul and started the drive up again.

On the drive up, I began to think about how empty the box truck I rented was and how it seemed like there wasn't too much stuff I had loaded up from storage. I started running the items through my head, when suddenly I thought, MY TV! Where was my TV? I reached for my phone and called my mom.

"Mom, what did you do with my TV? I don't think I loaded it up in the back."

"Oh," she said quietly. "I think I threw it away."

"You did what?" I couldn't believe how quickly and casually she had answered my question.

She said maybe she gave it to one of the guys who helped her get rid of some of my stuff, but more than likely, she loaded it onto the trash trailer and took it to the city dump.

The TV had a vertical line running through it about two inches thick, but it wasn't black and wasn't too bad—just a minor inconvenience. She could've given that TV away to someone or something, but instead, she just threw it away.

I was pretty mad but hung up as calmly as I could and kept driving. Then I thought, where are the girls' toy boxes? So I called my mom back less than an hour later.

"Oh," she said again. "I gave those away to the guy that helped. Did you want those?"

I almost lost my mind. I yelled and screamed and was just completely blown away that she had given away almost all my stuff. I rented a U-Haul twice as big as it needed to be, and then to add insult to injury, I had to spend the money to rebuy things that she threw out. She claimed ignorance, but she could've easily called me anytime and asked me about what I wanted to keep. Instead, she was mad at the situation and did it in haste with no regard for our stuff.

CHAPTER THIRTY-SEVEN

THE POWER OF OUR MINDS

I finished the trip and finally made it back to our apartment, but no one was around to help me unload or ground guide me in the big U-Haul. I started just throwing stuff around, not giving a fuck because I was mad at the entire situation. I was pissed that I was living in an apartment, pissed that I just drove 2,500 miles for essentially nothing, spent nearly five thousand dollars with everything associated. I was just done with it all.

When I arrived from Texas, we spent all day for weeks continuing to apply for jobs. The holidays came and went quickly, and we spent most of them locked up in our apartment, seeking employment. When I arrived from Texas, we spent weeks applying for jobs. It was depressing.

During the holiday season in 2016, I saw something on Facebook by The Seattle Times about their annual "Ignite Education Lab" and started to read more about it. They wanted people to call in and leave a voicemail about a story they have on how education helped them in their lives, so I called and told the story of Sly saving me in high school.

2017 came, and we continued our job search. We would wake up, cook breakfast and eat, apply for jobs, make lunch and eat, apply for jobs, make dinner and eat, then apply for jobs or play some videogames online friends and Conner. I started drinking daily to replace cannabis as I was trying to get clean if a drug test was required for a potential job. It was horrible, but I didn't enjoy sobriety much, so I started drinking. I was gaining weight at an exponential rate and was becoming something that wasn't me.

I remember looking up at myself in the mirror one of those days and seeing just how much I was a shadow of myself. I didn't feel like myself, and the interviewers likely saw that. I figured the two in-person interviews I had for okay jobs probably didn't hire me for this reason.

I was trapped; I didn't know what to do. As the months went by, we ran out of money, and I started talking to Army recruiters again. I determined that I would join the Army before I ever went back to Texas and looked at my options about re-enlisting.

Before making any offers, though, I decided to start the non-profit organization I had thought about doing for many years. Jack and Conner agreed to come aboard as co-directors. I wanted to make sure that there weren't any laws I would break unintentionally, so I spent hours on the phone with various agencies to ask basic questions. I started making the calls and doing what was necessary to figure out what I needed to do.

I talked with local Walmart managers about setting up outside the door to solicit funds. We wanted to start raising money for a lawyer because we needed to file for tax-exempt status with the IRS.

I was so excited; things were moving forward. I thought things could only get better from there, but I couldn't have been more wrong. I messaged Jack to let him know that I got the okay from Walmart to set the table up, and he told me he couldn't because he was starting coding school in Seattle.

I was happy for him because he used the GI Bill to go to school. In the end, that's all I want for my fellow combat brothers. At the same time, though, I was sad because I had been doing all this work, and he never mentioned anything about school in Seattle.

I fell into a more depressed state than before and just kept applying for jobs and jumped back into the same old depressing routine. I hit the bottle daily and hated everything about every day.

Then, I was five months into job seeking. After I connected with a veteran job recruiter online, a fortune five company finally reached out to Celina with interest, so she started jumping through their hoops. The process took over two months, and the minute they emailed her with interest, I stopped drinking, stopped applying for jobs, and started using cannabis again. I did this because I knew she would get the job; she's been offered just about every job that has interviewed her. I was so relieved that she got the interview and eventual job with that company, but as with most things that happened to us in the past few months, there was a shitty lining.

As she got the job, we were just about out of money, and I sold all our stock to float us to the end. Thankfully, the position came with a $20,000 sign-on bonus which helped us move out of the apartment and into a house.

It was surreal getting high in my new residence the first few days, and I was thankful to have found a fairly priced place compared to the market. I'm grateful that I had a house with a nice backyard for my daughters to play in and for Celina's good job. But as I sat on the couch every day, halfheartedly sending out a few applications and doing barely anything else, I felt like I was stagnating and stressing out at home.

Our new place was farther north of Seattle, making Celina's commute even longer, usually around three to four hours back-and-forth daily. She could use public transportation, and her employer paid the bill, but even when she drove to work, she still had about the same commute time.

The entire year she worked there, I felt like a single dad because she left for work before we were up for the day and wasn't home until after I made dinner. Things got hard for me; neither of us had ever experienced being a "single" parent, and the fact that I wasn't doing anything to engage my mind sent me into an accelerated downward spiral. I spent the last decade with my foot on the pedal, trying to load as much as possible. Then, suddenly, I was at home every day.

In one of my fits of restlessness, I decided I wanted to get into graduate school, so I started to look into different programs at the University of Washington. I found programs at Genome Science, Molecular and Cellular Biology, and The Institute For Protein Design and researched their departments.

One day, I was deep in my graduate program research when I realized that one of my daughters was starting to get sick because it was prime flu season. Amelia got over it pretty quick, but because Addie started kindergarten a few months prior, she was susceptible to getting extra sick, and that's what happened.

As I was caring for Addie, who was pretty much bedridden after a few days, I decided to apply to the Genome Science Department.

Their application deadline was less than six weeks away. I signed up and started studying for the GRE, but Addie's condition kept getting worse and worse.

She was coughing so horribly, and sometimes it seemed like she was on the brink of dying because of the pain she was in from an almost constant rough cough. We'd taken her to the doctors quite a few times, and they always wrote it off as influenza, but after seeing her cough on the couch like it was the last breath leaving her body, we retook her.

This time, her oxygen levels were low, and they took an x-ray of her chest. It showed the fluid building up in her little lungs: she had pneumonia. The doctor said that she was close to being hospitalized and that we brought her in just in time to get better with a generous antibiotic regimen. Those days were especially rough with my baby so sick, and I was as broken as a man could be.

The GRE, of course, didn't go so well, and when the deadline came, I didn't apply to the program I had my eye on because even had it gone well, it wasn't realistic that I would be able to go to school with Celina gone at least twelve hours a day. I didn't know what to do; I felt lost, depressed, sad, angry—just about every negative emotion and thing we could muster and feel.

***

The holidays came, and things were still grey and cloudy for me, and I did my best to put up a front for my daughters. I tried being "cheery," but I was an asshole most days, and any small thing would set me off. But before New Year's Eve of 2017, I sat down where I usually got high and did precisely that. I don't know what it was, but there was something inside of me that snapped, and that's the only way I can describe it. I sat there, and the thought came out of nowhere: 2017 had been one of the worst years of my life; I need to make 2018 one of the best.

At that moment, I didn't know how to do it. All I knew was I needed to change my mindset instead of waking up every morning with negativity. I would start looking at things positively, and I would begin to engage my mind more by doing something. I sat there and knew that I had to change my mindset because that was the biggest reason for the depression I was experiencing.

CHAPTER THIRTY-EIGHT

TIDAL CHANGES

2018 arrived, and five days into the year, I got an email from *the Seattle Times*. They said they enjoyed my submission from 2016 and offered me the opportunity to speak in February. It had taken such a long time, but I think fate knew I would need the pick-me-up in that very moment.

I had five weeks to prepare my talk and attend a few mandatory coaching sessions. It was a great experience, and I was grateful to share the experience with such amazing people. This book is about my life and not just my 5th brigade experience because the Seattle Times chose me. I practiced a lot over the next few weeks, made my slide show, and did the talk, which was quite the experience.

My first major public speaking event was an experience I'm forever grateful for

After the event, I was a bit scared about the aftermath of giving such a talk—probably more nervous than having to do it. I was frightened because I knew there wouldn't be another magical email that would give me something to work towards, and I knew I had to come up with something to keep my mind busy.

I decided that I would work on my minivan, and I looked online to see if I could find a tool lot, and I did. Pretty much everything I've used for all the work I've done to that vehicle has come from that $90 tool lot I purchased.

I also bought an air compressor and air ratchet to make some of the jobs more manageable, and eventually, my dad sent me his heavy-duty impact wrench. Without it, there was one job that would've been exponentially harder to do because I had eleven high-torque bolts to drop on the undercarriage frame. I completely redid the brake system, from new rotors to the replacement of the fluid reservoir. I upgraded the shocks and struts to aftermarket and changed the inner and outer tie rod ends to help suspension and steering. Part of that was changing the lower control arms, which required the impact wrench. Lastly, I finally worked on the exhaust system by changing the catalytic converter to get rid of the check engine light. Today that van rides smooth as a Cadillac: no check engine light, with better gas mileage, all at a total cost of around $1,100 for everything it took to do the work. I would buy them with damaged packaging discounts online to save extra money on the aftermarket parts.

Doing all that work on weekends encompassed a lot of time, and during the week, I was looking for deals on parts and watching YouTube videos on how to do the jobs. Then, when the van was all squared away, I had to find a different sort of project, so I picked up sewing because a lot of my petite wife's clothes needed altering.

Once again, I hit online deals on a sewing machine and picked up a pretty decent one for $40. After some more online video coaching, I employed what I learned and did a pretty decent job on the first project: a pair of pants that needed hemming. I did break some stitches initially and had to start from scratch, but overall hemming pants are effortless to do, and now it seems crazy to pay someone to do something so simple.

Shortly after that, though, Celina hit me with the real task! Celina came home one day with two oversized shirts that they gave her at work, and she told me, "Here you go, I need you to turn one of these into a size small for me."

I didn't think it would be easy, but I quickly learned otherwise after a quick online search. A video showed how to do it, and following along, I did it, and it came out better than I could've imagined ever doing.

We learned that one of our combat brothers who'd been battling a rare cancer was on his deathbed while going through this sewing adventure. Many of the men from 3rd platoon went to be by his side. I often saw his posts on how he was feeling on social media, which is where I initially found out about his diagnosis.

It was heartbreaking to see him so sick, and I couldn't help but think of all the stuff we went through to deploy and end up overseas where they gave us poison to drink.

A few days later, I played out back with my daughters because it was a beautiful summer day. When I glanced at my phone, I got word from one of my buddies that he had passed.

This message lit such a fire under me, and I decided to write about our experience in a short op-ed in the hopes of getting it published and bringing awareness to the issue. I went inside, got the computer on the bar, and wrote the piece in about fifteen minutes while standing. It was full of passion, controversy, rage, but most of all, truth. Then, at the request of a few people I sent it to, I worked on it some more.

I researched and found recent articles about water contamination on bases in the U.S. with a chemical used to fight aircraft fires. I thought it was a crazy coincidence and finished my work. Then I reached out to the author of those articles seeking guidance, and since it was the inspiration to start the book, I'll share what I wrote with you now.

> Recently, the World Health Organization launched a health review into bottled water after investigation and analysis by Orb Media discovered more than 90% of the bottles tested had microplastics in them[1]. However, this is water coming off store shelves that have been in mostly controlled environments and away from the sun. A report released in March 2018 by the House Armed Service Committee stated that there was water contamination at least 126 military installations across the US by a chemical known to cause cancer and birth defects that are used in foam to put out aircraft fires[2]. There are many atrocities I'm going to speak about in my life that I've noticed, and this one is going to be the first I try to bring to light as it affects soldiers from all over the world as opposed to just the men I served with. Water is a precious resource all over the world, especially so in war zones and I'll never forget my first experience getting it there. I deployed in July 2009 as a Sniper with 5th brigade 2nd infantry division out of Ft. Lewis, Washington. We deployed to Southern Afghanistan and the first thing you'll notice at Kandahar Airfield is the wonderful stench of the poo pond, which is a smell you never forget as you're constantly greeted with it as you exit the chow hall. The water most troops have access to are

pallets with cases of bottled water exposed to the sun. My mind was blown at the fact that we spend trillions of dollar on the war, but can't spend a couple thousand to build adequate shade for the pallets to keep them from being beat down by the sun. The waters were horrific to drink as you could taste the chemicals that seep into it that leech from the plastics. I wasted a lot of water going through many to find one that was actually potable, which took more effort than I'd like to admit. I knew that it wasn't good for anyone in 2009, but now after getting my degrees in chemical engineering and chemistry I realize the extent and severity of the problem. The worst part about all this is that it's a ticking time bomb with various lengths of fuse, and there will be a lot of people getting, if not already, sick from this. The immune system takes a serious blow in the months leading up to deployment and during. From the anthrax and smallpox immunizations (as well as others), burn pits that release toxic gas, to malaria pills that they now know cause brain damage, to the only readily available water source being filled with molecules that mimic hormones like estrogen. I'm working now to get water samples from overseas to test and see what kind of chemicals are leeching and their concentration, and after more research, I've learned that there was a recall in 2004 that was on the same water we consumed in 2009[3]. It's hard to know exactly the impact this issue is going to have as the concentrations are likely much higher from direct sunlight exposure. This is an issue that is hurting not only our soldiers, but everyone who drinks them and my only hope by bringing awareness to this, is that we can enact some change because it would take minimal cost and effort and have a profound effect on the health and readiness of our troops and allies.

1.https://orbmedia.org/stories/plus-plastic/

2.https://www.militarytimes.com/news/your-military/2018/04/26/dod-126-bases-report-water-contaminants-harmful-to-infant-development-tied-to-cancers/

3.https://www.stripes.com/news/85-000-cases-of-water-recalled-in-afghanistan-1.22082

I emailed the woman who was writing the articles to ask if she knew someone who could help me do something about the water quality overseas, and I also sent her my work so she could see that I was serious about the issue.

It was a weekend, and she responded from her iPhone, saying she liked what I'd been doing. She asked if I wanted to run it in their publication. I informed her I would be happy if they ran the article but that I wanted to submit it to *the Seattle Times* first because I wanted a civilian audience to see what I had written.

That week, after *the Seattle Times* declined the piece, I emailed her to let her know that I was ready to run it, but she never responded to my email. I considered emailing her back again, but I knew what more than likely occurred, and this is how I envision it went down.

I know she liked what I wrote because she confirmed that with me via email, but on that Monday, when she went to work and showed her superior my piece, they dug a bit deeper into 5th brigade to see what I would have to say about "atrocities" I saw while serving. Once they fell into that rabbit hole, they ran for the emergency box, broke the glass, and smashed the "Do not give a platform" button on me.

This was a situation I often imagined that year. On June 8th, 2018, we lost our brother in arms at twenty-nine years old. He was given up to two years to live but fought like a warrior and underwent intensive chemotherapy sessions to try and beat one of the rarest cancers known to man. He was and still is an inspiration to many, and I hope one day I can bring awareness to the potential harms that could be caused by the water the world's troops had to drink; it's the least they deserve. At this point, I was so motivated to write about my entire life experience and journey to reach a kid like myself in the hopes of just lifting one person.

THE STRANGEST THING HAPPENED when I started writing the book and recalling my earliest memories. I could feel the energy being zapped from my body as I sat at my kitchen table, typing away on my laptop. It hurt to remember all the things I endured as a child, especially because my daughters were the ages I was when I went through so much, and I couldn't imagine putting them through that. I shed a lot of tears and was working on it at a feverish pace until that bitch crept in.

Yeah, we all know who she is, and she is the reason that most of us sit on our laurels and don't strive to do more with our lives and the precious time we have. SELF DOUBT! She crept in hard and handed me the script.

"Who am I to write a book?" I asked myself. "I'm nobody. Everyone goes through trauma."

It's not even that I cared if anyone read it; I didn't feel like it was something I deserved. I am no one special, just a traumatized individual trying to navigate his way through this crazy system established by the people that came before me.

After a few days of thinking it over, I'd made up my mind. I wasn't going to write anymore. I wasn't sure how to keep from going crazy now that I had worked on every project I could think of: all the mechanic work, sewing projects, and writing were all finished.

We were in the fall of 2018, and I knew I needed something big to launch me on a course that would change my life forever. So, as you might have guessed, I tried to find some new project to turn to.

Again, I hit the think tank and stewed in a cauldron of ideas and paths I wanted my life to take.

I decided to go all-in and patent my idea from 2015, researching patents and looking up patent attorneys. I did a lot of digging through online forums and was thinking about filing the provisional patent myself to save money and give myself something to do.

The more I researched, the more that seemed like a bad idea. I found a patent attorney online who engaged with a user online, letting him know that he could use an attorney from anywhere in the country and not just a local source. I fell into this attorney's rabbit hole and liked his story because he worked hard to succeed in his law firm and career. The last thing I wanted was a patent attorney who didn't know what it was like to take a risk or have to push their life chips "all in" with something, and he had. But he was also one of the best in the country, and I didn't think he would deal with me. Because again, self-doubt crept in.

I contacted a patent attorney in Texas, one in Florida, then finally, after a few days, I set up a consultation with the attorney I eventually chose. I submitted the forms and everything he wanted online and even scanned some quick sketches of the invention for him because I misplaced the ones I did previously. I hadn't done that with the other attorneys, nor did I tell them anything about my idea.

I set up the initial phone consultation, and after reading over his firm's website, I was beginning to get a bit nervous. The site said the consultation could take fifteen minutes if the idea weren't patentable. On the other hand, if the idea was patentable, you should expect thirty minutes of discussion. I wasn't sure how the patent process worked, but I trusted that John wouldn't pull a fast one on me, so I laid all my cards on the table for him to see.

When the day arrived for the call, I eagerly anticipated what he would say. Of all my ideas, this idea had the biggest investment opportunity because it was cheapest to patent, simply because it didn't have any moving parts or electronics, and the market was a multi-billion-dollar-a-year space.

That was when he called. Shortly after he said hello, he got right into his thoughts.

"My two biggest concerns," he said. "Who you have told about your idea and that you used online patent search sites to search for prior art."

If his biggest concern was who I'd told, then there was nothing to worry about at all. I knew it had a lot of potential, so the only people I shared it with were people I thought could be interested enough to go in with me and put up some of the cost of the patent.

After the call, I had to wait a few days to hear about the course of action and its cost for a prior art search. At that time, I debated doing it because $1,500 would've been a lot of money to spend on a patent search that I might have easily done by myself. Unfortunately, like always, no one I approached was interested.

We sold our last shares of Celina's stock that she was given as part of her compensation package to fund the search and went on with the process.

While the prior art search was going on, I received my patent attorney's book that he wrote about his journey to starting his firm, and it inspired me to continue writing this book. We have to tell our stories because no matter how insignificant or "un-special" we may feel, someone out there may hear our message and see that ANYONE is capable of doing more for themselves with just hard work and determination. I got back on the writing horse and started going at it again, and this time I attacked it hard. This time, I felt as if I couldn't stop.

CHAPTER THIRTY-NINE

FROM ANOTHER MULTIVERSE

Over the winter and holiday season of 2018, I stopped writing for a while to indulge myself by spending extra time with my family and showering them with love. I did spend time preparing things for the book but not writing it, and the time spent was minimal, but I was just amazed how my entire life experience had turned 180 degrees with just a change in my mindset.

I still struggle even today, of course, but now I know that it's just a struggle, rather than cloudy weeks or months and uber depression.

Over this break, I spent a lot of time researching veteran entrepreneurial programs to see what I could apply to and ended up finding a great program that had a rolling application called Entrepreneurial Bootcamp for Veterans, or EBV. I applied because they offered online workshops and a curriculum on entrepreneurship, culminating in the cohort flying out to universities around the country for a week-long industry-specific workshop. I found other programs, but EBV seemed like the most involved since there was an in-person portion instead of everything being online.

I also made accounts through the Small Business Administration to take their online webinars. I wanted to start building my knowledge base. I also decided to read thirty books, bringing it down from fifty after a rational discussion with my wife. I was doing anything and everything I could to learn about business and what it takes to build and run one successfully.

2019 came, and I started knocking out books, webinars, and online videos because one of the applications I filled out asked for a "value proposition." I had no idea what that was, so I looked it up to see what I could find, like many things I didn't know. I ended up finding a ninety-minute video posted by an ivy league business school on the concept and over 200 videos on business.

I was doing everything I could to learn about starting and sustaining a successful business because I didn't see selling or licensing the patent being very likely. I was grinding away on this book while waiting for the patent to be finished and learning about business on most days.

While searching for entrepreneur programs online, I came across a Master of Science in Entrepreneurship program offered by Foster Business School at the University of Washington. The more I looked into it, the more it seemed like an excellent opportunity to forge my patent into the golden ticket I envisioned it to be. It was a year-long program, which is great because I was eligible for a year of tuition through the VA's vocational rehabilitation program. Even if I didn't qualify for that benefit, though, I would pay out of pocket for the course. Luckily for my wallet, UW halved tuition since I served in Washington. I decided that the investment was worth the opportunity to learn from some of the Seattle area's best entrepreneurs.

I applied to the program and felt I had a great chance of getting in simply because of my patent. I imagined how gratifying it would be to get in because of an idea that I came up with, all by myself, and then taking that idea and making something of it. I intended the patent to be a resumé booster and use it as a reason to network with people, as much as I suck at it. It's just hard being a massive introvert, though it may not seem like I am. I'd much rather be alone most days. This is probably because my experience with people most of my life has generally been shitty, so I have issues and know that I need to get over them.

When I started using my vocational rehabilitation benefits, I made sure to have all my ducks in a row in preparation for my first meeting with a local VA counselor. That was the first in a handful of hoops that I had to get through before being awarded the benefit, so I wanted to make sure everything was good to go.

I showed up a little early for my appointment, which I usually try to do because I hate being late. As I sat in the lobby, a couple of others came in and sat to wait for everything to begin. They eventually called us back, and we all sat together in a sort of "briefing" room to get a quick spiel on what we could expect, and then they had us watch an informational video that was about an hour long. After that was over, we waited in the lobby for them to call us back to meet with our counselors.

When I got called back, I came ready with a folder in hand to show my counselor everything I had in the pipeline. We began talking, and during the first thirty minutes of our meeting, we mainly discussed my military experience and hardships.

As I talked, I took a quick look around the office and noticed a degree posted on the wall. My counselor had his master's in psychiatry, and it clicked that he wasn't just a vocational rehab "counselor," but an actual mental health expert. That's when I realized that there was a reason we were talking about my hardships. Talking about military hardships was the easy part of the whole process, but discussing my recent difficulties made the meeting harder.

I told him about my patent and how I hoped to get into the entrepreneurship program to learn more about the business world. We talked a bit about it, and then I went on my way, unsure of what would come in terms of getting vocational rehabilitation and getting into the master's program.

A couple of weeks went by, and I eventually got an email from the entrepreneurship program stating that I was placed on standby and would need to do another interview. I was then worried a bit because in my statement, as part of the application process, I talked about how I wasn't a businessman but a scientist and engineer who needed to learn about business.

The next couple of weeks were hard because I wasn't sure what would come from doing this second interview. As part of the voc rehab process, I had to get things done by the veteran's office at the University. After mulling it over for a bit, I decided one weekend that I would take a trip to the university to get the paperwork done. I also decided to take my daughters with me to get a small experience of walking a large university campus. Hell, it wasn't something that I had experienced, so I wanted to share the experience with them.

I planned to go the following week on that Monday, but something happened on Sunday before that made me want to change plans for the day. But I was sure going to go sometime that week, and as it turned out, it happened to be that Wednesday when I decided to go.

I had spent the last few weeks hyping myself up. I didn't want to be afraid to step outside my comfort zone anymore. I knew getting uncomfortable was the secret to success. I wanted to brainwash myself into being okay with uncomfortable situations. With that, I told myself that when I went to do the paperwork for voc rehab, I would also stop by the offices of those in charge of the entrepreneurship program to say hello.

I was tired of being a shadow of myself, and what I knew I could be. I just needed an opportunity, and the opportunity only came if I dove headfirst into the mouth of the discomfort devil. Even though I'll always naturally want solitude, I know I'll never get it. With that, I also know it's something that drives people insane, and many people say they wish they could be alone. It's something we say we want, but is it? I feel like I've experienced small amounts of it, and I think something good comes with that. As a child on a ranch with no internet, cable tv, and no one to interact with, I was able to spend a lot of my life reflecting on what was going on around me and the things "adults" were teaching me. But now, all I wanted was for them to see me in person to hopefully influence a positive decision by allowing me into the program.

My daughters and I took the trip that Wednesday, and it didn't take long to get the paperwork signed at the veteran's office. As we rode the elevator down to the first floor, anxiety started to build. When the ride eventually stopped and the doors opened, we walked and exited the glass doors of the building, and I stood there, heart pounding in my chest. All the hyping up I did the days and weeks prior had quickly faded on that elevator ride, and I stood there scared out of my mind. It wasn't the thought of stopping by that frightened me; it was the thought of the upcoming interview hurting my chances of getting in.

I didn't want to feel like I was imposing myself on the decision, and sliding down that rabbit hole took my confidence with it. I stood there for what felt like days while my daughters pulled at my arms, eagerly wanting to explore the campus. I fought hard with myself in my mind— my mental warfare game was strong that day. Suddenly, I was a little boy again, being asked by the state trooper, "Son, who do you want to go home with, your mom or your dad?" The anxiety overflowed in my body.

And just like then, I decided for the better and went forward with what I had initially planned. I stabbed the anxiety in the face, grabbed my daughters by their hands, and began walking towards the Foster Business School Building.

CHAPTER FOURTY

IMPOSTER IN ME

After a short walk, we arrived at the building, and I swallowed hard before walking into the office area. As I walked in, I saw the director and another person talking.

"RICARDO!" the director said, almost surprised as I walked in. "I thought I just finished interviewing you."

I was lost for a second, but then she went on to clarify that she had just finished going through my application packet and watching my *Seattle Times* talk. Then, it was apparent on her face that she had an idea.

"Do you have time for another interview now?" she asked.

"If you don't mind my daughters sitting here and drawing while we talk," I responded.

"Absolutely not!" she said. "They're fine sitting and drawing right here," as she gestured towards the student studying area.

I quickly set them up to try and entertain themselves for a bit, and all the while, my heart was pounding so hard that I could feel it in my temple. I tried to calm myself when walking by, thinking of how Ricky Ricardo would tell Lucy she had some " 'splainin to do." I did my best to feel in the same situation as Lucy.

As I mentioned before, when I applied to this program, I told them I didn't want to be a businessman because I genuinely didn't want to be. I just wanted help learning how to turn my patent into a golden ticket to help me fund a biotech company with its profits. I didn't know about the grants available to life science startups at the time, so my focus was on turning the patent into something that would fund another business. That way, my brother from another universe and I could try to do something meaningful for humanity in our lifetimes.

After about forty-five minutes, our meeting concluded—my daughters, who were fantastic with only one slight distraction, packed up their little backpacks. We left, and I couldn't help but feel like I had punched my ticket into the program. I was so glad that I took the "Mamba" mentality and went for the big shot. I've been a huge Kobe Bryant fan my whole life. It was his influence that helped convince me to walk into that office. I'll never forget the interview he did not long before he passed. He mentioned pitching his idea for "Dear Basketball," the award-winning short video on basketball and its influence on his life that ultimately won him an Oscar. He said when he pitched the idea to the person capable of helping him make it happen, they told him, "Come on man, give your elevator pitch. You got thirty seconds."

Wow, I thought. *Standing there was one of the most famous people on earth trying to pitch an idea, and even he was forced into having to do it in less than a minute.* This put things in major perspective for me.

If Kobe wasn't afforded the time to fully explain his idea, who am I to think I'll get any time allotted from someone's day to voice my ideas? This realization was quite the reality check, and it drained a lot of the entrepreneurial soul from me. But I still fought through because

Mamba Mentality is all about stepping up when the spotlight is the brightest, and the road is the roughest.

As the weeks went by, I eagerly anticipated the email that would determine my entire future. I wanted so badly to get in, and when the day finally came, it was surreal. I got an email stating that I got into the program, and it finally felt like the waves of shit that had been thrown at me my entire life were starting to recede. I couldn't believe that I would be a student at Foster Business School, one of the best in the country. I, a "scientist" and a kid who came from so little, went to business school!

The entire experience in the program was worth the imposter syndrome and the four-to-six-hour commute. I couldn't do a lot of the "after school" activities as I had to be home by a certain time to pick up my daughters from elementary school. But still, it was one of the best experiences of my life.

Orientation for the program was a couple of days, and we were catered to with fantastic food and drinks, alcoholic and otherwise. Even hard liquor was available, and I opted for my regular whiskey on ice and enjoyed the environment.

As I caught a buzz, I walked away from the crowd and out of Dempsey Hall, Foster Business School's main building, drink in hand to enjoy the view of the campus. I walked out of the building, but only just a few feet from the doors as I just wanted to overlook the open area in front of the building. As I looked out at the campus and sipped on my beverage, the sights were so much to take in. There wasn't a person around, which made it all so surreal. I was lost in it all; after everything I've been through and everything I've done, here I was.

This all felt like a dream like I was tapped into that other universe the same way I had during my time before being sent home during Special Forces Assessment and Selection. Except this time, it didn't feel as real and was more alcohol-induced. And since I hadn't drunk regularly for several months, the hard liquor gave me quite the buzz. I relished it for as long as I could, and when someone finally broke the solitude I was enjoying, I went back inside the building to join my peers.

As I finished my drink, I took one last look at the campus and turned to walk back into the building. I was about to walk into what I hoped would be the beginning of my new road, one with fewer potholes and dead ends. Will it turn out that way? Only time will tell. All I knew at that moment was that it was time to grind away and find a way!

Acknowledgments

I started writing while deployed to Southern Afghanistan in 2009, and it's been a long journey to finally get this done! I want to thank Anna Ciummo, a fellow Husky, for taking the leap to help me edit this book. Going to the English Department for help was one of the best things I did during my time in business school. She could've dumped the project when the coronavirus hit, but she patiently waited and worked as I went through the transition as well.

RB, without your help, getting this thing to the finish line would've been a more painful and time-consuming process. After our first meeting, I was relieved to know that I have a friend who understands book publishing and mentorship. I'm so thankful to have met you and even more thankful for your willingness to help veterans and give back in the multitude of ways that you do. I hope we can continue collaborating and continue pushing for positive change for our communities.

Sister, I know things have been hard for us, and writing this book opened my mind to a lot of things I never thought about before. You've been through so much yourself, and I'm proud of you for what you've done thus far. I know you have it within you to do whatever you want in life, and I'll be here to support you when you're ready to make those sacrifices.

Mom and Dad, you've used and abused me for as long as I can remember. Thanks for always keeping some form of food and shelter available to me, buying me plastic blocks and videos games, and for trying your best to parent when you occasionally made an effort. It still blows my mind how the two of you were too blind to see the tiny golden ticket that you all created. I had so much potential and youth stolen from me, fixing that party blowout toy as a toddler was just the tip of the iceberg.

Sly, you're such an incredible human being. I craved the plant knowledge from such a young age. To think you were 5 miles away from me doing that exact thing for children in our town, I probably would've become the Albert Einstein of plants by my senior year. I still hope to take you on that badass vacation Slick, sorry it's taken longer than expected, but that's on par for my life. I never could've imagined not being able to find work after graduating in 2016, but I grinded through it all to end up at the tip of the spear in the fight against COVID. Thanks again to you and your wife for coming to my graduation, I wish I could've spent more time with you all!

Monica's family, there is nothing I feel I could ever to do properly thank you for the years you spent essentially raising me. It wasn't very long in the grand scheme of things, but I was with you just about all day, everyday and it hurt when I wasn't anymore. I felt like I lost a family, and when you have none, things truly feel empty.

Julian and his family, thank you for making me a part of your extended family. There are so many times you fed and housed me, and are a big reason I could stay out of major trouble in my early youth. Julian, from the earliest days of beating Oregon Trail on the old Macintosh computers, to today, we've always been good friends. Take care of yourself so we can enjoy the good life when the cards finally fall in our favor.

Thank you to all the other families in Freer, Texas, and the rest of the South that opened their doors to me. Without the small help from the culmination of you all, things would've been much different.

Eddie, Sergeant ED! When we were all there as new privates in the ether without weapons or assignments, I was hoping to be an SDM with my experiences growing up. When they put me on the 240B, I was pretty upset, I think I even tried to get out of it because I knew where I could be of best use. But it stuck, and I became grateful. Though carrying that 30lb monster was tough at first, it, along with your leadership, helped prepare me for the sniper tryout. I always gave you my all, and I appreciate you being there for me at the end. You knew that me being a "malingerer" was outside who I am, and I know you're voice was limited as a Staff Sergeant but you spoke in my defense anyway. You're the pinnacle of leadership, and the Army is lucky to have kept you for as long as they have!

Clint, making the sniper team with you and Stephen was literally a dream come true. To be a fresh recruit, private first class, teamed up with two Ranger tabbed combat experienced warriors was more than I could've ever dreamed up. From you bunking next to me on main post when we had nothing as a unit, to venting in your CHU overseas for 8 hours almost 3 years later, I guess your influence was just bound to be. You're an incredible warrior, and I'd gladly go to war with you again without hesitation.

Dennis, my OG Sniper buddy! Dennis, you like many of us lower enlisted, arean incredible leader and soldier. They ran us into the ground at Pre-Sniper, and I hadn't quite become the strong runner I later morphed into but you got me started on the track. That three weeks of everyday running was definitely the start of that journey, but while I was there sucking, you were already such a beast. You never complained once about it, and I always pushed to be as strong a runner as you. There is no doubt in my mind that we were the best sniper team at Pre-Sniper, I wish we had the records of all bullseye's we hit! I'll always appreciate you trying to talk me into staying in the Army right after you got back from overseas, it let me know that you still respected me as a soldier.

Chester, BIG RIG! Go DEEP! Haha man Chester you were so poopy about that football, but man it was such a hit! You're a good man Chester, having you in the platoon kept things interesting, both on and off duty. You letting me have access to your apartment before deployment was such a blessing, and I'm thankful to have served alongside you.

Tracey and Terrence, you crazy motherfuckers. Kept me on my toes with your expansive knowledge and were some of the few people that I could chat with about some deep shit. We took the plunge together for that coveted green beret, and ultimately none of us ended up with it. I'm always here for you guys! I know shits been rough since the days of our "youth," but I'm always here.

Axel! You were a great soldier and are a great man! Seeing you push for classes to work on your ASVAB score and actually getting to go to a few was inspiring. Being able to vent with you in garrison about all the stupid bullshit was some of the best times I had in the

Army, and serving with you overseas was by far one of my biggest honors, especially in light of you recently told me.

Jack! I have to admit, and I'm pretty sure I've told you this, but when you took the sergeant spot and kept me from being promoted, I was poopy. But after learning about your experience, and realizing that even if I made sergeant the other sergeants wouldn't respect me as such, I'm thankful that you came in. You were a new face with more time in service than a lot of those with more rank than you, so they gave you a thousand-fold the respect they would've ever given me. Though you're the reason my shoulder blew out, I'm thankful to have had the opportunity to learn and grow more as a soldier. Just keep in mind that we aren't all made of steel like you, so be cautious when giving people workout advice.

The C. Co Sniper Team, Man, you guys were something else. You all were the closest-knit sniper team in the brigade, and being able to spend all the time I did with you all in your CHU overseas was something I'll never forget.

The "Crew" There are so many of you that I spent time with on and off duty that I'd love to individually acknowledge, and it hurts me that we don't talk like we used to. Things felt like they really changed after I went to selection and subsequently got hurt. I wish the inept Army docs could've properly diagnosed me and got me into surgery right away because the 4 month recovery time would've fallen almost perfectly in line with when I deployed anyway. I would've been able to serve in a much better capacity overseas and since we didn't get out to the FOB until pretty much September, there was more than enough time to recover from surgery had I got it shortly after being injured.

My academic journey would've never been the same without:

Tacoma Community College! You set me up for success coming out of the Army, and I'll always shout your name from the rooftops. Without you and the MARC, I would've never become the algebra beast that I was. No doubt in my mind that you're one of the best community colleges in America!

Dr. Emi, thank you so much for being such a great professor. Taking General Chemistry II with you my first semester back in school after an almost year break was tough. Then when you asked if anyone wanted to volunteer in your lab the following semester, I couldn't believe the opportunity and of course had to jump in. I learned a lot in your lab over the years and appreciate you offering your support in pursuing my chemistry degree when I wanted to switch to just chemical engineering. The extra year I spent to get the chemistry degree was worth the effort!

The Engineering Department, I appreciate you offering me the scholarship opportunity as a veteran undergraduate engineering student. The money went a long way to help my family, especially towards the end when the financial aid department gave me trouble for the entire semester. It was fun to learn more about the electron microscope and how scientists characterize nanoparticles, and to learn that we had an electron microscope on campus was mind blowing!

Thanks to the thousands of jobs I applied to who didn't give me the time. This likely would've never got done had I been the one with a living wage job as opposed to my wife.

The Seattle Times, I was in such a deep and ugly pit in 2017 and for over a year things got just about as dark for me as my days in the military. This time the hurt was different but felt worse because now I had two babies that I was dragging through the mud along with me. Your opportunity may not seem like much to others, but to me it was exactly the hand I needed to start pulling me out of my depression.

The MS Entre program, thank you for all the work you've done to create such an incredible program. I wish I could've done more, but having to pick my daughters up from school in the afternoons made things difficult. My wife and I gave it our all though, and I'm forever grateful for being given the opportunity to learn from incredible people in such an amazing environment.

Cindy, Thank you so much for taking me on as a client and doing the cover for the book! To think that I would connect with someone who was familiar with the unit was just another mind bending part of this entire saga. I thought getting a response on Twitter and potentially getting the cover done by Goggins cover artist was going to be the craziest part of the process. I'm thankful she was too busy with her memoir to help me because your connection to the unit made it that much more special!

Finally, a big thanks to you! 100% of the revenue from this go back towards sharing the stories of the men who served along side me. Their stories are way more incredible than mine, so your contribution helps bring those to light. If you'd like to follow the journey in getting the documentary done and see how the money is being spent, be sure to subscribe to my newsletter on my website RicardoPPerez.com.

I'd also be in your gratitude forever if you'd leave a review, whether it be positive or negative! I'd love to hear what you hated or what moved you, so please share!

Thank you for your time, I hope you feel it wasn't a waste!

About The Author

Ricardo Perez served as one of the first and last soldiers in an Army Brigade that no longer exists. After a grueling sixty-five-hour tryout, he was selected to serve as a sniper alongside two other soldiers with special operations and combat experience. Though his military career didn't go as planned, he took the discipline learned to obtain degrees in Chemical Engineering and Chemistry while running a small business as a Texas State Handgun Instructor and doing research. The Seattle Times invited him to share a story at their "Ignite Education Lab" event, and he's a graduate of The University Of Washington's Master of Science in Entrepreneurship program offered by Foster Business School. Currently, he's a Research Scientist on the Next Generation Sequencing Team helping in their wartime effort in testing and sequencing the novel coronavirus at The University of Washington Virology department. He lives with his wife and kids in the Seattle, Washington area.